Derwentcote Steel Furnace

An industrial monument in County Durham

D Cranstone

With contributions by

M Chard
G McDonnell

1997

Published by
Lancaster University Archaeological Unit
Storey Institute
Meeting House Lane
Lancaster
LA1 1TH
(Phone: 01524 848666; *Fax:* 01524 848606)
(E-Mail: archunit@lancaster.ac.uk
(Internet —World Wide Web URL: http://www.lancs.ac.uk/users/archaeol/unit/luau.htm)

Distributed by
Oxbow Books
Park End Place
Oxford
OX1 1HN
(Phone: 01865-241249; *Fax:* 01865-794449)

Printed by
Kent Valley Colour Printers, Kendal, Cumbria

ISBN 1-86220-011-4
ISSN 1354-5205

Copy editor
Martin Lister
Indexer
Susan M Vaughan
Design
R Middleton
Layout and production
R A Parkin

This report is published with the aid of a grant from the Historic Buildings and Monuments Commission for England (English Heritage).

Front Cover: The restored Derwentcote Steel Furnace (© English Heritage)

Lancaster Imprints is the publication series of Lancaster University Archaeological Unit. The series covers work on major excavations and surveys of all periods undertaken by the Unit and associated organisations.

Contents

List of illustrations

Figures

Plates *(Chapters 1, 3, and plates section)*

Tables

Abbreviations

DRO   Durham County Record Office
DUDPD   Durham University Department of Palaeography and Diplomatic (now Durham University
              Library, Archives & Special Collections)
LUAU   Lancaster University Archaeological Unit
NRO   Northumberland County Record Office
OS   Ordnance Survey
RCHME   Royal Commission on Historic Monuments of England
TWAS   Tyne and Wear Archive Service

Contributors

David Cranstone
LUAU, Bolbec Hall, Newcastle upon Tyne, NE1 1SE

Margaret Chard
Jaramswic, Sand Croft, Penrith, Cumbria, CA11 8BB

Gerry McDonnell
Department of Archaeological Sciences, University of Bradford, Richmond Road, Bradford BD7 1DP

Acknowledgements

The programme was undertaken on behalf of English Heritage, funding being in the form of grants to the Tyne and Wear Industrial Monuments Trust (TWIMT). I am grateful to David Sherlock (Inspector of Ancient Monuments, English Heritage) and Ian Ayris (TWIMT) for their support and overall management of the project. I am also grateful to Alun Griffiths (Area Superintendent of Works), Hilton Higginbottom, Stephen Mansbridge, and John Whitfield of English Heritage for provision of equipment and other practical assistance, and to Alan Moody of Douglass Wise and Partners (consultant architect) for continued close liaison during the site-consolidation programme. The excavation team consisted at various times of David Gale and Kevin Haynes (assistant supervisors), Jane Buchanan, Susan Hedley, and Jenny Vaughan (finds assistants), Peter Buchanan, Ian Fernyheugh, Antony Francis, Chris Richardson, Alan Rushworth, Graham Stobbs, and Mike Smurthwaite (archaeological workers/site assistants), and Hazel Fleming and Matthew Watson (volunteers). I am grateful to all the above for their hard, cheerful, and intelligent work, sometimes under unpleasant conditions. Rectified photography was undertaken by Rob Burns (then English Heritage guardianship archaeologist for the region) and John Stanger.

Archive drawing was undertaken by Margaret Finch and David Gale. Office space was provided at various times by Dr Stafford Linsley (Centre for Continuing Education, Newcastle University), Colm O'Brien (the Archaeological Practice, Newcastle University), Paul Bidwell (Tyne and Wear Museum Service), and The Cranstone Consultancy. Other assistance was provided by Dene Conway, Tessa Grey, Anne Liddon, and Martin Wilson. Specialist reports were prepared by Edgar Bolton, Margaret Chard, Louisa Gidney (Durham University), and Gerry McDonnell (Bradford University). Sandra Hooper prepared the finds catalogues, and Jenny Vaughan advised on some aspects of interpretation. Drawings for publication were prepared by Margaret Finch (finds), Sandra Hooper (finds), Antony Francis (site plans and elevations), and Iain Hedley and Ruth Parkin (final amendments). I am grateful to Dr Stafford Linsley and Mr David Sherlock for commenting on a draft of this report.

I am also grateful to John Harrison, Peter White, and Drs J K Almond, Ken Barraclough, David Crossley, and Stafford Linsley, for their advice and comments on visits to the site, and must especially single out the late Ken Barraclough (whose encyclopaedic knowledge of the historical steel industry will be sorely missed following his recent death) and Stafford Linsley (for support, advice, and local knowledge throughout the project). I am also grateful to the Chapman, Giles, and Hutchinson families of Derwentcote for their local knowledge and for allowing me to inspect their homes.

Historical research was undertaken in the Borthwick Institute for Historical Research (University of York), Bradford District Archives, Cumbria Record Office (Carlisle branch), Doncaster Archives Department, Durham City Library, Durham Record Office, Durham University Department of Palaeography and Diplomatic (now Durham University Library Archives and Special Collections), Durham University Library, Gateshead Library, the India Office Library and Records, the Literary and Philosophical Society of Newcastle upon Tyne, Newcastle Central Library, Newcastle University Library, Northumberland Record Office, Nottinghamshire Record Office, the Public Record Office (Chancery Lane and Kew), Sheffield Record Office, and Tyne and Wear Archive Service. In addition, Mike Spencer undertook research on my behalf in Derbyshire Record Office, and correspondence with several other repositories, companies, and individuals obviated the need for personal visits. I am grateful to all the above for their assistance.

I am also grateful to all the colleagues and friends who have helped during the historical research, or commented on drafts of the Historical Archive: Brian Awty, Ian Ayris, Frank Atkinson, and John Gall (Beamish Museum), Harry Beamish (National Trust), Dr Eric Clavering, Dr J B Nolan of Cookson Industrial Materials Ltd (for permission to use company archives in Tyne and Wear Archives), Dr Chris Evans (Newcastle University), Ian Forbes, Peter King, Grace McCombie, Philip Riden (College of Cardiff, University of Wales), Dr David Rowe (Newcastle University), Roger Sims (Tyne and Wear Archives), and Janet Swabey. Again I must single out the late Ken Barraclough (for supplying unpublished extracts from Torsten Berg's translation of Angerstein's Swedish manuscript), and also Stafford Linsley for supervision and advice throughout the research, Chris Evans for supplying several obscure and important references that I would not otherwise have located, Peter Brown for undertaking additional research in the PRO (Chancery Lane), and Peter King for elucidating the history of the Gateshead steel furnace.

Summary

A forge at Derwentcote, Co Durham (NZ 130565), was built in *c* 1719, and operated until 1891. The cementation steel furnace which is the main subject of this report was probably added in *c* 1733, and remained in use until between 1875 and 1891. The works was operated by a series of relatively short-lived partnerships in the eighteenth century; from the 1790s to 1872 it was controlled by successive members of the Cookson family, and from 1872 to final closure in 1891 it was again operated by a series of short-lived partnerships and companies.

Externally, the furnace consists of a buttressed rectangular structure, passing up into a conical chimney. Internally, it contains an ashpit at the base, above which a firegrate opens into the base of a rectangular vaulted chamber containing the stone cementation chests. Flames from the fire were carried by a series of flues round these chests, and exhausted by a second series of flues into the base of the hollow chimney. The stone external structure is original, but most of the internal linings represent repairs and replacements during the working life of the furnace.

The interiors and surroundings of the buildings attached to the furnace have been excavated, in advance of public display. Within the Southern Building, this excavation revealed a series of floor levels, with evidence for smithing and for a charcoal-grinding mill. Outside the standing buildings, excavation revealed the sites of one paved yard and three timber buildings (two of which had been walled with brick panels infilling a timber frame, whereas the third had been entirely of timber construction), together with stratigraphic sequences relating to the construction, use, and abandonment of the furnace.

The large finds assemblage was derived largely from post-abandonment contexts, and was dominated in bulk by firebrick and process residues. The former included an interesting series of stamps, and the latter forms a type-assemblage for process residues from the cementation process. The ironwork assemblage included fittings from various parts of the furnace.

The furnace has now been conserved and opened to the public by English Heritage.

1

INTRODUCTION

Derwentcote steel furnace (NZ 130565) is a unique survival of an almost-complete cementation furnace from the early stages of the Industrial Revolution, displaying one of the technologies on which this revolution was based, and as such is one of Britain's most important industrial monuments. The importance of the site was recognised from at least the 1960s (Harrison 1969) and efforts by various bodies and individuals to ensure its preservation culminated in the site being bought by the Department of the Environment and placed in the guardianship of English Heritage in 1985. The structure has since been the subject of a programme of research, building recording, and excavation, of which this report is the outcome, and of a programme of consolidation and repair by English Heritage Directly Employed Labour and the outside contractors J and W Lowry Ltd (who provided a new roof), all preparatory to its opening to the public in 1991 (Plate 1).

The site lies in the modern parish and district of Derwentside, Co Durham, some ten miles south-west of Newcastle upon Tyne, on the south side of the river Derwent. The setting and layout are described below,

Plate 1   Photograph of site as conserved (© English Heritage)

but it should be noted at this stage that the site contains two spatially and functionally discrete units: the steel furnace itself (to the south-west, on higher ground), and the forge (to the north-east, close to river level). The former has been the focus of excavation, recording and conservation to date, whereas on-site work on the forge has been limited to tree clearance and landscape survey. This is reflected in the ensuing report: Chapters 2, 3, and 4 describe respectively the technological background, the physical setting and layout, and the history, of the site as a whole, whereas Chapters 5, 6, and 7 concentrate on the structure and the excavated archaeology of the furnace area in isolation. The final discussion (*Ch 8*) considers both units, but is inevitably heavily biassed towards the furnace.

The research programme

The immediate areas of the furnace and forge were taken into care by English Heritage in 1985, and the adjacent woodland in 1991. A programme of detailed recording of the standing structures on the furnace area, and excavation of the interiors and immediate surroundings, was conducted in 1987 and 1988, under the direction of David Cranstone. The bulk of this fieldwork was conducted in two seasons: September–December 1987 and April–June 1988. Further field recording was undertaken in the summer of 1989 (including a rapid structural record of Derwentcote House and Farm, the inhabited buildings to the southwest of the furnace), and a 'watching brief' was maintained during drainage works in 1989 and 1991, and consolidation of the standing structures in 1990–91, a small excavation was undertaken in 1991 to clear a site for the custodian's hut. The main programme of historical research was undertaken in 1989, with some additional work in 1990–91, partly as a thesis for the Local History Certificate of the University of Newcastle upon Tyne (Cranstone 1989).

Detailed typescript archive reports on all the above work have been prepared and lodged in the Site Archive, and contain the detailed evidence summarised in Chapters 4, 5, and 6; these are referred to below in the form Archive Report 1 etc. The Site Archive and finds have been deposited with English Heritage, and will be kept in their Helmsley store.

In addition, separate drawn surveys were undertaken on the exterior surfaces of the cone of the furnace (by computer photogrammetry, under the direction of Mr Ross Dallas of the Photogrammetric Unit of the Institute of Advanced Architectural Studies at York University), on the interior spaces of the furnace by Mr Colin Bamber and Mr Bill Blake of English Heritage drawing office, and of the surrounding landscape by Mark Bowden and Keith Blood of the Royal Commission on the Historical Monuments of England (Newcastle upon Tyne office). Information from these surveys has been included in this report.

2

TECHNOLOGY

The detailed metallurgical information obtained by examination of the site is described later (*Ch 6*), but the wider technological setting should be summarised at this stage; this section is derived largely from Barraclough's masterly review of the early steel industry (1984a and b, especially 1984a, 1–14).

In post-medieval Britain, iron could be produced in two forms:

1	Wrought iron, a commercially-pure product containing streaks of slag, which was malleable but relatively soft; it could be forged, heat-welded, and bent, but not cast (due largely to its very high melting point of 1535°C). Wrought iron could be produced directly from iron ore by smelting in a bloomery, where the iron did not liquefy, or by a process of decarburising ('fining') from cast iron, in a finery forge. The former process was universal until the late fifteenth century; by the early eighteenth century (when Derwentcote was built) the last bloomery forges were being converted to finery operation (Fell 1908, 202–3, 206; Crossley 1990, 154–6). The finery forge produced wrought iron in the form of plain bars ('bar iron'), suitable for trading and further forgings into finished goods.

2	Cast iron, an iron-carbon alloy containing 3–4% carbon, and melting at 1130°C (a rather higher temperature was necessary in practice), produced by smelting iron ore in a blast furnace and running out the molten product to solidify as 'pigs'. Cast iron could be used to make castings (either directly from the blast furnace or by re-melting in a foundry), but the vast majority of British production was used as the raw material for conversion into wrought iron in the finery forge.

Steel, as known before the nineteenth century, was an iron-carbon alloy with a controlled carbon content of 0.5–1.5%. It could be forged (with some difficulty), and its hardness, toughness and ductility could then be closely controlled by quenching (abrupt cooling from red heat) and tempering (gradual reheating to a lower temperature). Its precise properties depended on the carbon content, and were also very sensitive to impurities; any trace of sulphur or phosphorus rendered it unworkable, since it became too hard and brittle to forge. Most British iron ores contain phosphorus, and some contain sulphur. Since this passed into the metal, and could not be removed with eighteenth century technology, British iron was not normally suitable for steelmaking. In practice, Swedish bar (*ie* wrought) iron (from very specific forges, identified by their stamps on the bars) was the almost universal raw material for steelmaking from the early eighteenth century to the mid-nineteenth century.

Steel could be produced in various ways:-

1	Directly from iron ore by bloomery smelting under conditions which introduced some carbon into the iron.

2	From cast iron by controlled partial decarbur-isation. In the sixteenth–seventeenth centuries, this was undertaken in the finery forge, but the process appears to have been abandoned as cementation (*see below*) was perfected. The principle was however revived in the nineteenth century, using 'puddling' in a reverberatory furnace (the process which had replaced fining for the conversion of cast to wrought iron), and was briefly of major importance (Barraclough 1984b, 91–106).

3	By heating wrought iron in contact with cast iron, so that carbon diffused from the latter to the former. This might be done below the melting point of cast iron (the 'Chinese process') or above ('co-fusion', or 'Brescian steel'), the wrought iron/steel however remaining solid in both cases.

4	By heating wrought iron in contact with charcoal, in a reducing atmosphere. On the small scale, applied to previously-forged artefacts, this was 'case hardening', and the process was of considerable antiquity. On the large scale, using bars of wrought iron sealed in chests in a specially-constructed furnace, it was known as 'cementation'; the wrought iron bars were converted into bars of 'blister steel'. This process was apparently developed in Central Europe in the late sixteenth century; it was introduced into Britain in the early seventeenth century, initially with variable success. By the late seventeenth century, the process was becoming widespread in Britain, and the hinterland of Newcastle was developing as the centre of the British steel industry. In the eighteenth century the British cementation steel industry dominated European production, and was known on the Continent as the 'English Method'; its centre only shifted from Tyneside to Sheffield in the later part of the century. The local history, as it affects Derwentcote, is discussed in more detail in Chapter 3.

The cementation process depended on the heating of iron bars for a long period in close contact with charcoal but completely sealed from the air. This was undertaken in chests of refractory sandstone or firebrick, in which the bars were packed round with charcoal powder (sometimes with other ingredients which were believed to facilitate the process), the chests being sealed with a layer of sand, or sand mixed with the 'swarf' from grinders' wheels; the former was normal on Tyneside, and the latter in Sheffield. The chests were permanent fixtures within a circular or rectangular reverberatory furnace, in which the flames and heat from a coal fire were carried by a flue system round the chests, reverberating down from a vault above them, and vented to the atmosphere by (typically) a conical chimney similar to that over a pottery or glass kiln.

The raw materials consumed by a cementation furnace were therefore bar iron (in the eighteenth century invariably Swedish and from a restricted range of forges in the Dannemora or Öregrund area of Sweden (Barraclough 1984a, 36, 173–5), though in the nineteenth century a wider range of non-phosphoric irons was used)), charcoal (the species of tree used was probably not important, since the requirements specified by many contemporary accounts are very inconsistent), sand or a sand-based mixture for sealing the chests, and coal or wood as a fuel (coal was universally used in Britain, whereas wood was the normal fuel in Sweden, necessitating changes to the detailed design of the furnace). In addition, a supply of refractory sandstone (for the chests) and fireclay (for the firegrate and for luting the internal structures) was necessary, since the chests and internal linings of the furnace needed relatively frequent renewal.

Cementation was a batch process; Barraclough (1984a, 40–42) quotes a detailed description of the working process at a local furnace in 1767. The furnace was loaded by the steelmaker and an assistant, by passing the bars through small holes in the ends of the furnace; eleven tons were used per firing. The bars were packed round with ground charcoal, and the top of the chest was then sealed with dry sand. This could be done in about six hours, access being gained by climbing in through the ashpit. The fire was then lit, and took about 15 hours to heat the chests to red heat. The firing was then continued for five days and nights, the fire being stoked and tended continuously from both ends of the firegrate. The furnace was then allowed to cool for six to ten days, before the steelmaker re-entered and unloaded the bars, now of blister steel, through the small holes. It was customary not to carry out more than twelve campaigns in a year, and the chests lasted for 18–24 campaigns.

Other descriptions inevitably differ in the details of times (often longer than in the above description), sealing material, and the use or otherwise of additional cementing materials in additon to charcoal. The hot and intensely dusty and claustrophobic atmosphere inside the furnace is rarely commented on by contemporary observers, but can be imagined (Barraclough 1984a, 239). The Derwentcote furnace is likely to have held 10–14 tons per firing (depending on the ratio of charcoal to iron inside the chests), indicating an annual capacity of 100–200 tons. To put this in a national context, in 1737 only 1000 tons of Swedish iron were converted into steel in the whole of Britain (Barraclough 1984a, 61).

The blister steel produced was not directly usable; it was brittle, and its carbon content was not homogenous. Its surface was also marked by the 'blisters' which gave it its name; these were produced by carbon monoxide gas, from the reaction of the migrating carbon with streaks of slag in the bar iron. The lower 'tempers' (grades of steel, as defined by average carbon content) were worked by hot forging (initially under a forge hammer, later in a rolling mill), before use for making springs and the cheaper cutlery. Higher 'tempers' (0.9–1.2% carbon) were bundled into 'faggotts' and forged down at red heat, sometimes repeatedly, to produce an almost-homogenous finely-laminated product known as 'shear steel'.

The forgeability of steel decreased with rising carbon content, and steels of above 1.2% carbon were rarely produced before the later eighteenth century.

However the range of uses for steel was greatly increased after *c* 1750, when Benjamin Huntsman in Sheffield perfected the 'crucible process' for melting blister steel, in a sealed ceramic crucible set in a small but intensely-hot coke furnace (before this date steel could not be melted, due to the difficulty of designing a furnace to reach the very high temperatures needed, and refractories to withstand these temperatures). For melting, blister steel of up to 1.6% carbon content was often preferred, due to its lower melting point and the increased hardness of the product. Crucible steel could of course be used for making cast steel goods, which had not previously been possible; it could also be forged, and had the advantage of being completely homogenous in composition (since the melting process eliminated the slag content of blister steel).

The crucible process also made it possible to produce cast steel directly by melting a mixture of wrought iron with carbon or cast iron. The former was not used during the relevant period, due to the impracticality of reaching a temperature sufficient to melt the wrought iron in the crucible furnace. The latter was very much easier in practice, since the cast iron would melt at a relatively low temperature to produce a 'bath' into which the wrought iron could dissolve. However a major problem was the lack of any suitable cast iron, sufficiently low in sulphur and phosphorus to produce a good steel; Sweden did not permit the export of cast iron. This ban was lifted in 1854, and from this date the direct production of steel in the crucible began to supercede the use of blister steel; the use of unmelted forged steel was already declining, and continued to do so. The cementation process was therefore obsolescent, but its product was preferred for some high-quality applications, and the last UK cementation furnace (Doncaster No 2 Furnace in Sheffield) did not close until 1952.

It should be noted that the meaning of the word 'steel' has broadened with time; both blister and crucible steel would now be considered as types of carbon steel (which is still produced, by different methods). In addition, the name is also used for two very different products: mild steel is an alloy of iron with around 0.2% carbon, and has replaced wrought iron (rather than carbon steel) for most purposes; and alloy steels, such as stainless steel, are alloys of iron with a range of other metallic elements, and have partially replaced carbon steel.

3

THE SETTING

Geology

The solid geology of the Derwentcote area consists of Coal Measures (Carboniferous) strata; these are overlain by a mantle of Pleistocene (presumed Devensian) drift deposits, conforming in outline to the modern topography (which is therefore of pre-Devensian origin). The following description is taken from Mills 1982 and Geological Survey of Great Britain 1982, which together form the most recent published coverage.

The Carboniferous strata are composed of Westphalian A (Lower Coal Measures) beds, consisting of rhythmic alternations of sandstone, siltstone, mudstone, shale, coal, and seatearth (fireclay), with occasional ironstone beds (more accurately described as shales with abundant ironstone nodules). The strata dip to the east at $1-2°$, but are interrupted by a major northwest-southeast fault (the 'Tantobie Dyke'), downthrowing by up to 60 m to the south-west, which crosses the Derwent near the mouth of Bairn's Gill, and produces local steep dips. To the south-west of the fault the Brockwell coal seam outcrops at surface in the sides of Owlett and Bairn's Gills, the outcrop continuing beneath drift along the side of the Derwent valley and passing almost directly beneath the furnace. The (lower) Victoria seam outcrops under alluvium to the north-west of the furnace, though most of its extent lies below valley floor level. The German Bands ironstone, which lies between these seams, must also be present close to the site, though it is not clear from the published evidence whether it would outcrop at surface or be concealed by drift. To the northeast of the fault slightly earlier beds are exposed, including a yellow sandstone exploited at Derwentcote Quarry (and at other sites downstream).

The drift ('Boulder Clay') of the area typically consists of stony clay, varying in colour from yellow through dull yellow and grey to pale grey; the stone consists largely of angular fragments, and pockets of sand and gravel are also present. The predominant composition at the furnace site was in fact a yellow sandy clay. Over the area in general the drift cover averages 3–8 m in thickness. However the nearest boreholes to Derwentcote (near West Derwentcote Farm to the west, and between Owlett Gill and Haggs Farm to the southeast) show thicknesses of c 18 m, the drift thickening to form a shelf along the south side of the Derwent valley. The scarp to the north and east of the furnace is cut entirely through drift, exposures of the Carboniferous beds being limited to small outcrops in Owlet and Bairn's Gills, the riverbank east of these, and the riverbank below West Derwentcote Farm (some 300 m west of the furnace).

Topography

Derwentcote lies 14 km ($8\frac{1}{2}$ miles) south-west of Newcastle upon Tyne, on the south side of the river Derwent (Fig 1); it is 10 km (6 miles) above the confluence of the Derwent with the Tyne at Swalwell, above which the Derwent has never been navigable. The Derwent valley averages 5 km (3 miles) wide and 200 m (600′) deep, its floor lying at around 50 m OD and its lips at between 200 m OD and 300 m OD. The slopes of the valley are moderate rather than steep, but are interrupted longitudinally by steep side valleys and (especially near the base of the slope) small but very sharply-incised gullies. A bifurcating pair of these latter, Bairn's and Owlet Gills, open onto the Derwent floodplain less than 200 m east of the furnace, and form a considerable barrier to east-west movement for some 300 m to the south.

The base of the main southern slope of the valley has been intermittently cut back by meanders of the river, forming local steep scarps. One of these dies out 300 m northwest of the furnace; another starts 50 m northwest of the furnace, and continues east past the mouth of Bairn's Gill and Derwentcote Quarry. The gap between these scarps forms an area of easy access to the Derwent and its floodplain, and to a ford across the river.

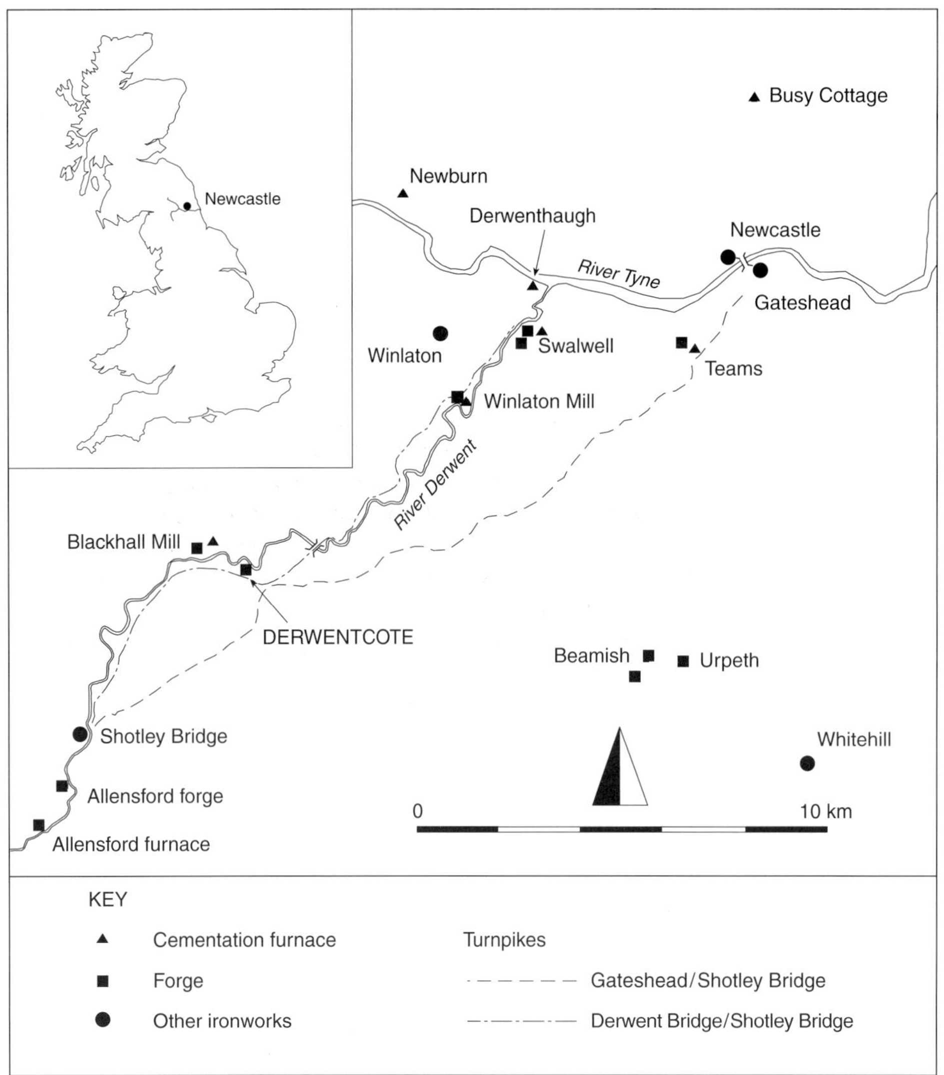

Fig 1 Derwentcote: site location showing other eighteenth and early-nineteenth century iron-and steelworks in the area

The forge lies on the floodplain, at around 50 m OD; its headrace, millpond, and tailrace occupy a slight natural hollow along the south side of the floodplain, along which a small stream flows. The forge and furnace are only 60 m apart but are separated by the wooded scarp, vehicle access skirting the lip of the scarp to the north-west of the furnace (along Forge Lane), and returning east along the floodplain. The furnace lies on the crest of the scarp, which curves from north-west to south around it; there is also a slight north-south hollow running down the overall valley slope just west of the furnace, which therefore occupies the end of a slight ridge, sloping up to the south.

Modern land-use of the area is varied; temporary pasture and arable dominate the agricultural use (the former perhaps predominating), with some surviving permanent pasture (mainly on the lands of Derwentcote Farm). The area is extensively wooded; almost all the steep slopes are tree-covered (largely by ancient woodland, modified to varying extents), and there are also large areas of woodland on the gentler slopes (predominantly plantations, both of conifers and of hardwoods; the plantations overlie both former agricultural land and ancient woodland sites, and some modified ancient woodland does survive).

The human landscape (Figs 1–2; Plate 2)

The immediate environs of the furnace and forge were surveyed by the Royal Commission on the Historical Monuments of England in 1990 (RCHME 1990). The following account is a summary of this survey, supplemented by study of the OS map coverage (1:2500 Plan, sheet V.15, various editions; *see also Ch 4*), and by the author's own observations.

The southernmost features of the industrial complex are the houses of Derwentcote House and Farm (separately described below), and Forge Lane running north from the A 694 road (formerly the Derwent Bridge / Shotley Bridge Turnpike, and constructed

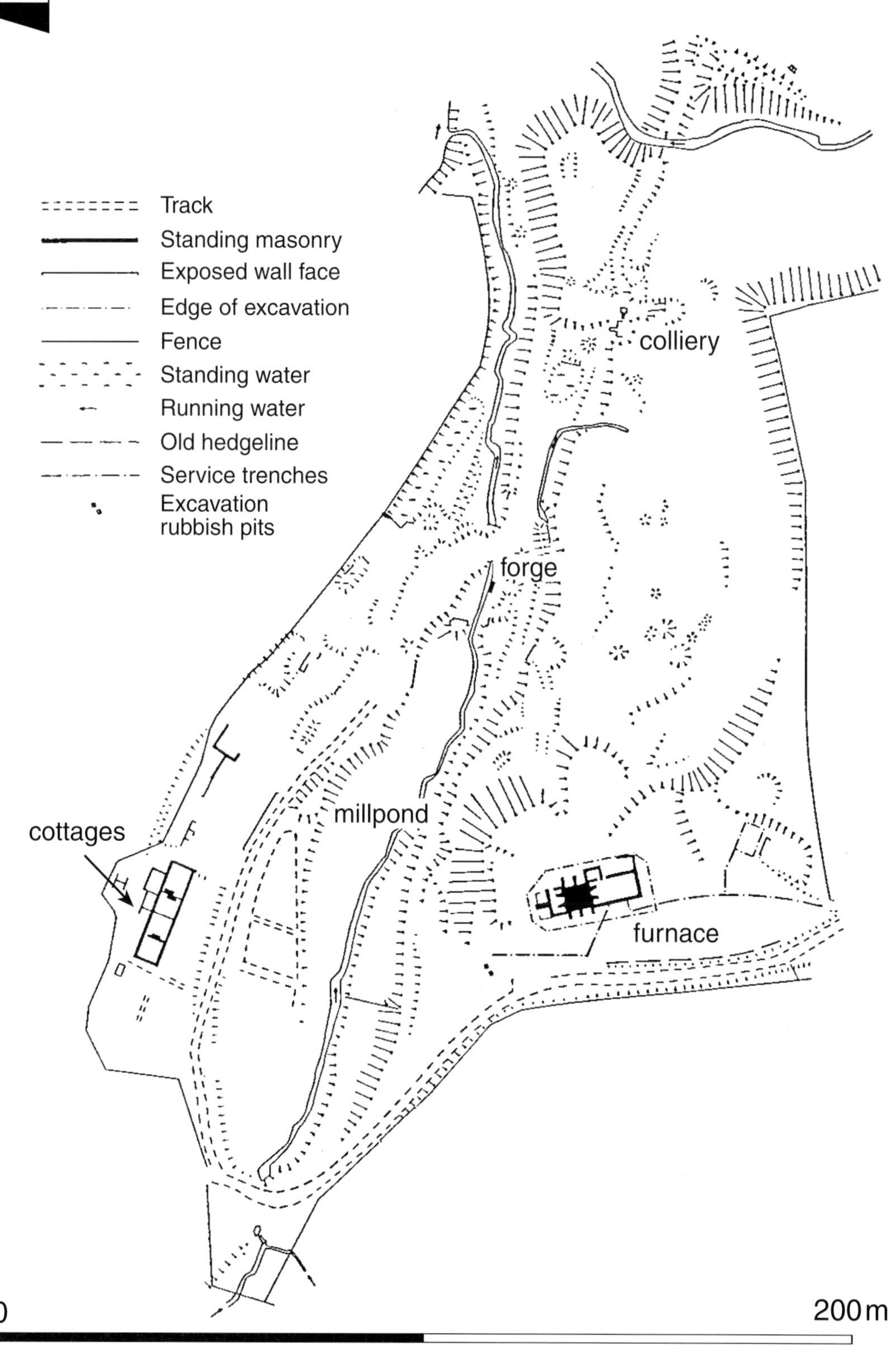

Fig 2   Overall site plan (redrawn with additions from a survey by RCHME)

on a new line in the 1830s (*see Ch 4*); the field to the east of this is improved pasture, with no visible evidence of former trackway lines across it. To the north of Derwentcote House, Forge Lane develops into a slight hollow way down the steeper slope, with further sub-parallel hollow ways in the (permanent pasture) field to its west. The west side of the present lane retains an intermittent wall of Type A slag blocks (*see Ch 7*). At the base of the slope, the Lane crosses the headrace, and continues across the floodplain on a slight causeway, curving north-west to the site of a ford across the Derwent.

Little trace of the dam or weir across the Derwent survives, but the headrace from this towards the forge survives as a prominent earthwork (up to 1.9 m deep), dry until it is joined by two small streams where it passes under Forge Lane. From here east, it merges into the silted and drained millpond of the forge (*see below*).

The remainder of the area, to the east of Forge Lane, is now occupied by woodland (of varying character), with only small clearings; the survey of it was inhibited by scrub in some areas.

The small area between the furnace and Forge Lane forms a clearing, cultivated until recent years (M Hutchinson pers comm). To the south-east of this, the line of a path from Forge Lane to the forge and the east side of the furnace is visible as a slight hollow way and terrace (the northward continuation of this, round the east and north sides of the furnace, has been located along the edge of the excavated area, but is not expressed as a clear surface feature). A bank beside the south end of this path divides it from a rectangular enclosure measuring *c* 25 x 9 m, terraced by up to 1.5 m into the slope to its east and south, and containing traces of a stone building associated with a scatter of slag Type A; the enclosure, but not the building, was present in 1856. To the east of this, a boggy hollow marks the 'well' shown in 1856.

The scarp along the east side of the enclosure connects at its north end with a major lynchet, up to 2.5 m high, curving from east to north-west towards the furnace, with a subsidiary lynchet diverging west from its crest near the north end of the enclosure, and traces of a trackway along its base. The main lynchet dies out locally south of the forge, but resumes as a feature up to 4 m high along the modern field boundary to the east, returning south along the edge of Bairn's Gill. These lynchets presumably form the remains of a field system (of unknown date, but probably pre-dating the furnace).

The area below the scarp, broadly above the forge, consists of confused topography on a steep northward slope. Many of the features in this area are probably the result of natural landslipping, but traces of deliberate terracing (perhaps for buildings or for cultivation) are visible, and others could be the traces of bell-pits or borrow-pits for mineral extraction. A particularly regular oval hollow, 3.5 m long and 1.2 m deep, is reminiscent in shape of a limekiln or calcining kiln, but there is no visible structural evidence to support this interpretation.

The confused area was bounded to the west in 1856 by the path from Forge Lane to the forge; the line is intermittently visible, but the more prominent feature on the ground is a track running directly from the forge to the furnace, as a hollow way and double lynchet, only the eastern part of which was used by the 1856 path line.

The upper (western) part of this track, approaching the furnace, forms a marked feature across the east end of an arc of flat-topped tips, radiating north and east from the furnace, and substantially modifying the natural slope of this area; the scarps of the tips are up to 4 m high. Where their composition is visible, the tips consist of assorted slags and metallurgical debris (all the common types from the excavation being represented; *see Ch 7*), together with building stone, tile, and slate.

Below these tips, the base of the steep natural slope is occupied by a well-marked double lynchet, indicating the line of a track running west from the south side of the forge to Forge Lane (in use in 1856), and rising gently to the west. There are scatters of metallurgical debris along its north side. Between this and the west end of the millpond, a north-south wall built of assorted slag masses marks the west end of a small enclosure shown on the 1856 OS map.

The millpond survives as an extensive marshy hollow traversed by a small stream, the dam having been breached in two places at its east end. It became narrower and shallower to the west, the west end not being precisely located in the modern topography.

To the north of this end of the millpond, the site of a group of gardens (in existence by 1856) is clearly marked by the survival of a hedgeline and internal stone-edged paths, with a coping of crucible lids at one point. To the north of this runs a partially-embanked track, forming the return of Forge Lane to the forge; this was the only vehicle access to the forge yard in 1856, but remained in use as the main access to the twentieth century collieries further east (*see below*).

To the north of the track, extending east to the site of the forge itself, lies a scatter of ruined buildings. The

Plate 2   Derwentcote in 1856 (taken from Ordnance Survey 25" map, 1st edition)

westernmost of these had apparently originated as a single rectangular stone building measuring 23 x 6 m, with large openings in its south wall (suggesting a non-domestic function). It had been converted into a row of three cottages before 1856, by the insertion of brick partition walls with ovens and fireplaces, and of extensions against the north side. A brick and concrete outhouse to the north dates from between 1914 and 1939 (from map evidence); it may have replaced a brick and stone outhouse to the east, abandoned over the same period. This latter outhouse backs onto a stone wall running east from the stone building described above, to a brick building 6 m wide by at least 6 m long; a building was present on this site by 1856, but was considerable altered later.

To the east of this building lies a large hollow (up to 1 m deep); this was water-filled in 1856, and presumably formed the west end of a channel along the north edge of the site (largely beneath the north range of the forge, and only partially visible today). A small brick ruin of late date lies to its south, and to the south of this a jumble of rubbble, wall bases, and scarps marks the site of the west end of the north

range of the forge buildings, as depicted in 1856 and 1895.

The track from Forge Lane originally terminated in the yard between the two ranges of buildings that made up the forge, only being extended eastwards between 1915 and 1939, when it was developed into the main access route to Forge Drift colliery (*see below*); this extension takes the form of a broad embankment running east south-east across the forge site and the (culverted) tailrace. The field remains of the forge are consequently incoherent.

Traces of the north wall of the north range (of brick and stone) survive along the modern field boundary. Immediately inside (south of) this are two lengths of west-east water channel, up to 5 m wide and 1.3 m deep, connected by a brick culvert; it is likely that the separate hollow to the west was part of the same system, though no connection is visible. The eastern aperture of the brick culvert marks the east end of the northen range of the forge; to the west of this the watercourse was within or under the building, while to the east it formed an open channel feeding into the

11

tailrace from the southern range. Slight traces of the south wall of the northern range are visible just to the south of the western part of the channel.

The site of the north-east corner of the southern range of the forge is marked by a cluster of earthworks immediately north of the twentieth century embanked trackway. The most prominent of these is a stone-revetted hollow some $3\,m^2$ and 1.3 m deep, close to the mapped position of the corner; this feature is not understood. A gully up to 1.3 m deep runs east from the site of the east wall of the range — this is likely to have been a tailrace, but was disused by 1856.

To the south of the embankment, visible remains of the south side of the forge are also fragmentary. Parts of the dam survive, breached in two places; between these breaches the outer face, forming part of the forge structure, survives as a stone wall 2 m high, with a return at the north end indicating the position of a wheelpit or channel (from which the disused tailrace just described may have emanated). Map and photographic evidence indicates that the southern breach is also on the site of a channel, feeding (presumably via a wheelpit) into a tailrace; the line of these is more-or-less followed by the present stream course, and a short length of stone walling along the south side of the stream may be a remnant of the wheelpit. Terraced-in platforms to the south of this look like building sites, but were outside the south side of the forge by 1856. The east-west track along the base of the natural slope continues just south of these, crossing an embanked and stone-revetted east-west channel, which can be identified from map evidence as the overflow channel from the millpond. The direct path from the forge to the furnace (*see above*) also crossed the overflow channel, one abutment of a bridge surviving.

Almost all the visible features to the east of the forge site are the products of twentieth century coalmining; much of the area has been surveyed, but is only very briefly described. The main workings are centred on NZ 132565, the site of Forge Drift. The site of the 'drift' (horizontal mine entrance into the hillside) is visible, with an adjacent group of ruined buildings and other structures. A large tip of colliery spoil extends north and north-east from this to Owlet Gill, over the site of two small enclosures (from map evidence). Other 'drift' entrances and associated spoil heaps and tracks are scattered along the sites of Owlet and Bairn's Gills. Map evidence indicates that the first of these drifts, and the track up Bairn's Gill, were constructed between 1895 and 1915. Forge Drift (a much more substantial working) came into use between 1915 and 1939, but most of its spoil tip dates from after 1939, although it is now well-wooded.

The only remaining feature of relevance to the iron- and steel-works lies c 150 m northeast of Forge Drift (beyond the limits of the RCHME survey). This is the remains of Derwent Cote Quarry, presumably the source of building stone for the industrial complex and other local buildings. The plan form of the quarry remains unaltered from the 1856 OS map, although it appears to have been still in use when this was prepared.

The built landscape

No building recording has been attempted on the ruins of the forge or of Forge Cottages, since (although some structural information is visible) no clearance has yet been undertaken, and the structures are obscured by rubble and vegetation.

The only surviving inhabited buildings at Derwentcote lie to the south and south-west of the furnace. The nomenclature of these buildings, and of others in the area, has changed over the past century, and remains ambiguous (*see Ch 4*). Of the three buildings, the Toll Cottage, at the junction of Forge Lane with the A 694 road (formerly the Derwent Bridge to Shotley Bridge turnpike), dates from the 1830s, and had no close functional or structural connection with the steel-works; no structural record has therefore been attempted. The remaining buildings, Derwentcote House and Derwentcote Farm, have functional connections or constructional similarities (respectively) to the furnace, and were therefore recorded (in 1989) as part of the project. The detailed record is presented in Archive Report 3, and is briefly summarised below.

Derwentcote House (NZ 13025641) lies some 80 m south of the furnace, on the west side of Forge Lane, and faces south. In its present form the building is broadly 'L'-shaped, the long axis running west-east with a northward return at the east end, both these elements being of two-storied gabled form. Much of the angle between these elements has been infilled with lower accretions. Five structural phases (D1 – D5) could be distinguished, but D4 and D5 post-date the 1st edition OS plan (surveyed 1856), and are not described here. All the walls had been altered and repaired over the years, and a recent heavy cement pointing had obscured much of the external structural detail. Internally, most structural evidence was inevitably obscured by plaster, paintwork etc.

The earliest phase (D1) survived as the west end of the main east-west range. It had originally been a rectangular two-storey gabled building measuring some 12 x 4 m, with chimneys at the apex of each gable. The walls were built of roughly-coursed sandstone masonry (irregular on the north wall), with quoined corners. The south wall contained a blocked doorway just west of centre, and two windows on each floor (all altered to varying degrees). On the west (gable) wall, the masonry construction terminated at a flat head, above which the (fossil) gable was formed of dark red brick in English bond, terminating in a brick chimney with projecting string course, which may or may not have been original. The eaves had been heightened by c 0.5 m, contemporary to the addition of D2; on the north and south walls the addition was of small sandstone masonry, whereas on the west gable it was of lighter red brick, which also formed the corners. Internally, the building had contained two rooms on each floor, and was of single-pile form.

The second phase (D2) formed the north-south range, with a slight westward return at the south end to connect it to D1, both elements being longitudinally gabled. It was built of roughly-coursed sandstone masonry, slightly darker than that of D1; the masonry of the north wall was very rough and varied in composition, while all corners were quoined (except where they abutted D1). The south wall contained the main doorway of the house, beside the corner of D1; there was a window over this, and a pair of windows central to the wall (one on each floor). The east wall contained two windows, while the north wall was totally blank. There were chimneys at the apices of the eastern and northern gables.

The roof of D1 and D2 formed a single construction, of uniform Welsh slate with tile ridge-pieces with plain close verges. The internal structure was not visible.

Phase D3 formed an outshut along most of the north side of D1, terminating against the north-south range of D2. It consisted of two stories, lower than those of D1 and D2. The north wall was obscured by later addition D4, and the roof was modern; the west wall was largely composed of light red brick, in English bond bedded in hard brown sandy mortar, resting on a low stone plinth.

Phases D1 – D3 had been built before 1856, when they are shown on the 1st Edition 1:2500 plan. D1 may well be of eighteenth century date (extensive replacement and obscuring of original features precludes close dating). D2 was probably conceived as an addition to D1 rather than as a discrete building; an earlier nineteenth century date would fit the rather undistinctive architecture.

The historical evidence (*Ch 4*) shows that in 1856 Derwentcote House was occupied by C E Cookson, the operator of the forge and furnace; it is almost certainly the building occupied by Cookson's 'agents' in 1841, 1851, and 1871, and by the 'steel manager' in 1881. It is well placed to monitor access to the works down Forge Lane. It is perhaps surprising that it does not show any detailed structural similarities to any of the buildings in the furnace complex.

Derwentcote Farm (NZ 12995641) lies to the west of Derwentcote House, on a similar axis but offset somewhat to the north. It is reached by a track from Forge Lane behind Derwentcote House; this now gives access to the north side of the farm, but formerly (from map evidence) turned through the gap between House and Farm, giving access to the south front of the latter. The farm formerly consisted of three main units, a farmhouse in the centre, with a stable or barn to its east and an outhouse to the west, forming a single east-west range. In 1988 the barn was demolished and the farmhouse partially rebuilt and extended to the north. It is unfortunate that recording was not undertaken until after this rebuilding; however I am grateful to Mrs Hutchinson for showing me photographs of the building before and during the alterations, which have allowed interpretation of some features that do not survive. Surviving non-modern walls were in general unaltered and clearly visible.

The farmhouse was a rectangular gabled building measuring c 10 x 5 m. Only the south and west walls had survived the 1988 rebuilding. The west wall was partly obscured by outhouses, and could therefore only be recorded from a distance. It rested on a stone plinth. Above this the main structure was of thin bricks (230 x 120 x 50 mm) laid in Flemish bond. A change from darker to less dark brick suggested a fossil gable line, with eaves c 2 m from ground level, and a stone chimney at its apex. This was overlain by a second fossil gable, with a very clear verge of 'tumbled-in' brickwork and plain brick chimney. Above this, the top 1 m of the gable was formed of sandstone rubble with some re-used brick, with a quoined corner to the south face; this quoined corner extended downwards almost to ground level, cutting through both fossil verges. A plain chimney of machine-made white brick overlay the upper of the fossil chimneys at the apex.

The south wall was composed of uncoursed sand-stone masonry bedded in lime mortar, of one build with the (greyer) sandstone quoins on the southwest corner. The elevation was symmetrical, with a central doorway and two windows on the ground floor, and three windows on the first floor. The base of the south-west corner was composed of dark red Flemish bond brickwork (continuous with that of the base of the west wall), to the east of which a fragment of walling in grey laminated sandstone bedded in soft sandy mortar survived, broken off and abutted by the main build. Both of these rested on a slight stone plinth (only exposed near the corner). The south-east corner had been rebuilt in 1988; photographic evidence of its former character is described below.

The north wall is known only from photographs. Its lower part was abutted by an outshut; the walls of this were composed of small sandstone masonry, and the roof of stone flags laid in diminishing courses, and rising from *c* 1.5 m high at the eaves, to *c* 4 m where it abutted the main building. There was a timber lean-to to its west. The lower part of the main north wall was of brick construction, with a flat top forming a fossil eaves line continuous with the 'tumbled-in' brick gable on the west wall; the darker brick of the earlier fossil roof-line was visible near the north-east corner, but the division could not be traced elsewhere. the brick wall contained doorways into the outshut and the lean-to, but the only window was above the roof of the lean-to, with no evidence for blocked windows behind the lean-to or outshut. The top of the brick walling corresponded in height to the roofline of the outshut; above this level the north wall was composed of sandstone rubble, rather rougher than that of the south wall, with no openings.

The roof of the farmhouse was gabled longitudinally, with chimneys at the apex of each gable. The roofing was of uniform grey Welsh slate, with plain close verges, and grey half-round ridgepieces (presumably of tile).

The wall between the farmhouse and the barn to its east had been destroyed in 1988, but photographs of its east face were available. The lowest 2 m (approximately) was a thick sandstone wall, whitewashed to the east and roughly quoined on its western corners; the north-west corner was abutted by the dark lower brickwork of the farmhouse north wall (the southwest corner may have been abutted by the stone south wall of the farmhouse, but no clear evidence was visible). Above a flat ledge, the stone walling was carried up as a thinner gable, with a stone chimney at its apex (although no fireplace was visible on the east face of the wall). This fossil gable was overlain by a second fossil gable, with verges of tumbled-in brickwork, with stopped corners to the east; to the north-west it was continuous with the lighter brick wall of the farmhouse, and to the south-west it returned behind the stone facing of the farmhouse south wall. There was a brick chimney at the apex. This gable in turn was overlain by a gable of rough partly-rendered sandstone masonry, extending to the pre-1988 verge of the farmhouse, with quoined east corners continuous with the sandstone masonry of the north and south walls of the farmhouse; on the south-east corner these continued downwards, abutting a broken end to the brick gable wall. The apex of this final phase was

surmounted by a plain chimney in machine-made firebrick.

No overall photographic record of the barn existed; it had been derelict for some time before demolition in 1988. The building had been rectangular, continuing the east-west axis of the farmhouse. The north and south walls were of sandstone, with no features in the western 2 m (the only parts visible on the photographs), and were bedded in sand (Mrs Hutchinson pers comm). The varying relationships to the farmhouse walls at the west end indicate that the upper parts of the barn walls must have been a heightening, but no direct evidence of this was visible in the very limited photographic evidence. The roof was pantiled, and was at the same level and pitch as the tumbled-in brick phase of the farmhouse gable.

At the west end of the farmhouse, the western outshut survived unaltered by the 1988 rebuilding. It measured *c* 5 x 5 m, and abutted the west wall of the farmhouse, though it was offset northwards from the axis of this. Despite its small size, its structural history was quite complex.

The bulk of the south wall was composed of un-coursed sandstone rubble with occasional bricks, bedded in yellow sand and pointed with white sandy mortar with pebbles. The south-west corner was quoined in rusticated ashlar, while the east end abutted the primary dark brick phase of the farmhouse. The wall had been heightened by *c* 1 m, with sandstone and machine-made firebrick bedded in cement. The west wall was very varied. It contained a doorway north of centre, the south side of which was quoined in similar rusticated ashlar to the south-west corner; the intervening masonry was bedded in sand and pointed with white mortar. To the north of the door, however, the wall was of sandstone rubble set in white sandy mortar, with no quoins on either the jamb or the north-west corner. Above the doorway, this masonry had a broken end, abutted by very rough masonry containing large cobbles; this extended to 1 m from the wallhead over the door, but was only three courses high to the south, where it was overlain by an area of assorted brick and firebrick walling. Both these builds were overlain by *c* 1 m of machine-made white firebrick walling, continuing the heightening of the south wall. The north wall was composed of uniform roughly-coursed sandstone rubble, heavily coated in cement. The roof was pitched to the west, with plain close verges; it was covered with asbestos tiles.

Structurally, the lower stage of the barn was the earliest of the known buildings. However the presence of a chimney in its western gable with (so far as can be ascertained) no fireplace below it on the inner face of the wall, coupled with the existence of a very similar stone chimney at the apex of the earliest (dark brick) phase of the west gable of the farmhouse, suggests that the ground floors of the barn and the farmhouse were in fact contemporaneous, the butt joint indicating merely the sequence of construction. The farmhouse was then heightened in two stages, in the second of which the south face was also refaced in stone, with a symmetrical facade. The barn was subsequently also heightened. The structural history of the western outshut is not fully understood.

The 1st edition OS 25 inch map (Plate 2) shows that Derwentcote Farm had attained its pre-1988 plan form by 1856, though some of the heightenings could of course be of later date. The general form of the architecture might suggest a late seventeenth-to-earlier eigtheenth century date for the first phase of the barn and the first two phases of the farmhouse, and a late eighteenth-to-early nineteenth century date for the third phase of the farmhouse with its symmetrical south elevation.

The use of sand bedding in the stable (though it is not certain which phase) and in the first phase of the western outshut forms a close parallel to the construction of the ancillary buildings of the furnace (*see below, Ch 5*). This probably indicates a continuing quirk of local vernacular building technique; it need not indicate a close similarity of construction date, though this is entirely possible.

4

THE HISTORY

The history of Derwentcote in the wider context of the steel industry has been discussed by Barraclough (1984a, especially 67–9). Considerable further research was undertaken by the author during the course of the project. The results were however disappointing; it was not possible to locate any surviving records of the estate on which the site lies, or of the businesses that operated it, nor any detailed map earlier than the First edition OS coverage (surveyed 1856). Consequently use was made of very varied, and in some cases indirect, sources. The research has been written up in detail in the site archive (as the History Archive (March 1990), revised by the Additional Notes (November 1990)), from which the following text is simplified with some additional research. Due to the oblique nature of the documentation, some of the interpretations depend on large numbers of documents, in various repositories, and on chains of argument from these. These arguments and references will be found in the Archive texts; the present text refers directly to primary documents only when information is directly derived from this document.

The setting and physical development

The River Derwent, as a large and fast-flowing river, formed the power source for a range of industries at various periods (Flinn 1955, E Clavering pers comm, S Linsley pers comm). As well as numerous corn mills along its length, these included (in geographical order working downstream): the lead industry of the upper Derwent Valley around Blanchland and Hunstanworth; further lead mining in the Muggleswick–Healeyfield area; a blast furnace and forge at Allensford (Linsley 1978); the Hollow Blade Sword Company at Shotley Bridge (Jenkins 1935), and later paperworks and large-scale cornmill; woollen and sawmills at Ebchester; the forge and paper mill at Blackhall Mill (Plate 3); Derwentcote; paper mills at Lintzford; a forge at Gibside; and Crowley's ironworks at Winlaton Mill (partly replacing a fulling mill; Cranstone 1991) and Swalwell.

In addition, the valley from Shotley Bridge to the Tyne is cut through the Coal Measures, and formed an important part of the Durham Coalfield. The development of coalmining is outside the scope of this report, but it may be briefly noted that large-scale mining developed in the eighteenth century, concentrating on areas whose topography allowed waggonway access to the Tyne (in practice the Lower Derwent valley and the higher slopes in the Derwentcote area) (Bennett *et al* 1990, Lewis 1970), becoming general throughout the coalfield in the nineteenth century with improved road and (particularly) rail transport.

The place-name 'Darwen Coat' (in various spellings) can be traced back at least to 1569 (Surtees 1820, 292): it lay within the chapelry of Medomsley in the parish of Lanchester, and had its own mill by 1660 (this may well have been on the same site as the later forge). Before the late eighteenth century, communication links appear to have been poor; the Derwent has never been navigible, and the waggonways followed the upper sides of the valley, over a mile from the site (Lewis 1970). There was no direct road link down the Derwent valley towards Newcastle, but eighteenth century county maps do indicate (diagrammatically) a route from Ebchester along the valley side south of Derwentcote, then across the high ground to Ravensworth (1 mile south of Teams, *see Fig 1*). This route may be marked by an alignment of hedgerows, passing 350 m south of the furnace. Various eighteenth and early nineteenth century estate plans (none of which includes the furnace itself) suggest that Forge Lane was part of a network of trackways along the valley, presumably giving reasonable access to Blackhall Mill and Shotley Bridge (via at least one ford of the Derwent).

Communications began to improve in the later eighteenth century, as the turnpike network developed (see Linsley 1992, 73 for a map of the network, with dates of Turnpike Acts). In 1793 the Lobley Hill (Gateshead) to Shotley Bridge Turnpike was authorised. This passed some 600 m south-east of the furnace, where it was joined by Forge Lane; its route in both directions involved considerable climbs onto high ground, though presumably the road was at least reasonably built and maintained. A further improvement came in the 1830s, when the Derwent Bridge to Shotley Bridge Turnpike was constructed; this provided a direct route along the Derwent Valley (the modern A 694) via a totally-new bridge and road from Linzford to Rowlands Gill, and passed 200 m south of the furnace (again on a new line, cutting across the previous field system (DRO: Q/D/P/47; NRO: 324/G4)).

Cartographic evidence for the physical development of the forge and furnace complex is very limited, since there is no detailed coverage before the first OS survey in 1856. In 1811, a map of the area immediately to the west (DRO: D/Bo/G/27 xiii) does not show the headrace from the Derwent to the forge, and its dam on the Derwent. This may be merely an omission, but it is possible that the forge millpond was initially supplied entirely by the stream that still flows through it, the headrace from the Derwent being added after 1811 to augment the water supply. The headrace and dam are first shown on the Medomsley Tithe Map of 1842 (copy in DUDPD; the map is too schematic to give useful information on other buildings).

The site is first shown in detail on the 1st edition OS 1:2500 plan (Durham sheet V.15; surveyed 1856, published 1862) (Plate 2). The furnace and its Northern and Southern Buildings and buttresses were in their surviving form, the Southeast Building being shown as an unroofed enclosure or yard. The area west of the forge contained several buildings and gardens, some of which (plot 44) survive as ruins. The west end of the forge (plot 49) may have been warehouses, beside the access track. To the east of this, the north side of the forge was formed by an irregular building (plot 47) with a watercourse beneath its north side. The south side was formed by a second irregular building (plot 48) with a wheelrace along the south side, and perhaps another emerging from the north-east corner; ruins of this building survive immediately east of the millpond dam.

The only known early photograph of the works (Beamish Museum, Negative 12663) (Plate 4) shows this same building (plot 48) from the west; it may date from the 1860s (John Gall, pers comm). The building was gabled east-west, its eaves lying close to the level of the dam crest; the roof was pantiled and sagged along its ridge, and the visible wall was largely weatherboarded. An extension to the north-west corner was stone-walled, and seemingly continuous with a capped-off chimney stack rising to the apex of the main building. A large free-standing brick chimney, with iron strapping, lay to the east of the plot 48 building; this might well have served a reverberatory furnace. Its top was badly damaged.

A long brick chimney, with at least four pots, rose from the north side of the building or nearby; the form is typical of the chimneys of a crucible melting shop.

By the 2nd edition OS plan (revised 1897), the furnace (though not the forge) was identified as disused; the weir on the Derwent had been rebuilt on a slightly different line, and slight alterations had occurred to the forge buildings. By the 3rd edition (revised 1915), however, the Derwent dam had gone, the millpond was drained, and most of the forge buildings were roofless (though their outlines survived); the cottages north-west of the forge survived, as of course did the furnace. Coalmining had commenced from Owlet Gill to the south-east. By the 4th edition (1939 revision), the forge had totally disappeared and an access track to the colliery had been driven across its site (probably by infilling, to judge by the modern topography). The main colliery was now Forge Drift, 80 m south-east of the forge site; the surviving spoil heap indicates that this expanded considerably after 1939.

Derwentcote: the eighteenth century

The development of the iron and steel industries of the Derwent Valley have been discussed elsewhere (Barraclough 1984a, Barraclough and Awty 1987, and History Archive). It should be noted here that a steelworks was set up in 1687 by Dennis Hayford at Blackhall Mill (1 mile west of Derwentcote), at least in part to supply the Hollow Blade Sword Company at Shotley Bridge; a cementation furnace at this site survived until 1916, and is shown in detail on several photographs, but need not date from the initial foundation of the works. The steelmaker at Blackhall Mill was William Bertram, a German who appears to have introduced the manufacture of shear (or 'German') steel to Britain, and maintained a monopoly within the country until after mid-century.

The first mention of an iron or steel works at Derwentcote comes in 1719, when Kalmeter's diary mentions a forge held by Alderman Reed; the absence of any mention of a furnace is probably significant (Barraclough 1984a, 67, and K Barraclough, pers comm). Alderman Ralph Reed of Newcastle died in 1720, he was an occasional customer of the South Yorkshire partnership of ironmasters, of which Hayford was a prominent member, and in September 1718 he bought four tons of cast-iron forge plates from Chapel furnace (Sheffield); these may well have been destined for Derwentcote, and may mark the construction of the forge. It would appear that Reed was in partnership with William and Richard Thomlinson (respectively the brother and nephew of his wife Isabel), operating ironworks at Derwentcote, Skinnerburn (Newcastle) and Teams (Gateshead), and that in 1721 Isabel sold her late husband's share in the business for £3000 to the Thomlinsons, who operated the works for several years under the name of Richard Thomlinson and Co (NRO ZAN M13/C7).

In May 1733, the lease of 'Derwent Coat Mill otherwise Byarside Mill', with lands adjoining, an iron forge, warehouses, workinghouses and other buildings, was sold by Richard and William Thomlinson to a partnership of Cuthbert Smith, Thomas Wasse, Ralph Harle, and George Blenkinsopp, who intended to set up a business for the sale, forging and manufacture of iron and steel (Blenkinsopp being the agent for the business); they also took on copyhold land at Shotley Bridge. In 1734 James Moncaster was added to this partnership, which had a stock value of £4000 (PRO C11/381/113). The partners agreed to sell the business in 1742, and a sale notice duly appeared in the *Newcastle Courant* (October 23rd 1742). The property consisted of 'Darwentcoate Forge, together with a Steel furnace, a Mill for making French Barley, Warehouses, Workmens Houses etc, all in very good Repair, except the Barley Mill', and had 75 years lease remaining; the copyhold estate at Shotley Bridge was also included in the sale. Enquiries were directed to Thomas Wasse, Merchant, in Newcastle, and Joshua Cotton was to shew the premises. However it failed to sell, and in May 1743 the majority of partners agreed to relieve Blenkinsopp of management. A complex legal dispute resulted, further complicated by the deaths of Thomas Wasse (intestate) in 1745, and of Ralph Harle in 1746; Blenkinsopp disputed the claims at length, though his replies (which shed little light on the development or operation to the site) appear to be aimed to prevaricate rather than to provide substantive answers to the accusations (PRO C11/564/21/25).

In 1748 both the *Newcastle Courant* and the *Newcastle Journal* advertised for sale:

> Darwentcoat Farm and Forge, a Steel Furnace, and Smith's Shop, Warehouses, and Workmen's Houses, and other Erections therewith enjoy'd, held by Lease for a long term of Years, and of which about 72 years are yet to come; with the necessary Materials and Utensils now used with the forge, with the Bellows and Anvil belonging to the Smith's Shop; together with a small farm, adjoining to the said Darwentcoat Farm, held by Lease for a term of Years yet in being, taken for the Convenience of Way-leave to the said Forge; which said premises are about eight Miles from Newcastle, and five from Swalwell; together with a Copyhold Estate at Shotley-bridge, with a Water Corn Mill, and several Dwelling-houses and Smiths Shops, lately belonging to the Sword-blade Company, about ten miles from Newcastle, and seven from Swalwell.
>
> NB. There is a Stock of Charcoal, old Iron, Wood, and other materials, upon the premises, to be disposed of to the Purchasor upon a Valuation; and also a set of Workmen to be contracted with, if thought proper; and the Conditions of Sale will be produced at the Time and Place above-mentioned.
>
> Mr Joshua Cotton, and Mr Joshua Copley, both at the Forge, will shew the Premises.

The sale was first advertised on 30th January, but was postponed on 13th February due to objections from George Blenkinsopp; clearly the legal dispute had not been resolved. The 1742 notice fits well with a foundation date for the forge of 1718 (assuming a 99 year lease); the discrepancy in the remaining term of the lease between this and the 1748 notice cannot be explained (unless part of the site was covered by a separate lease of 1721, after Ralph Reed's death).

These advertisements clearly show that the steel furnace was in operation by 1742; the fact that no steel furnace is mentioned in the 1733 sale (as recited in the PRO documentation) may imply that it was built by the new partnership in or after this year. The corn mill had clearly closed between 1742 and 1748, presumably incorporated into the forge. Blenkinsopp was manager of the Hollow Blade Sword Company in 1753 and 1758 (S Linsley, pers comm), and the copyhold land at Shotley Bridge suggests an earlier connection. It is also interesting to note that Joshua Copley was a son of William Cotton of The Haigh, a partner of Hayford's in the South Yorkshire ironworks (Awty 1957, 96); Hayford's steelworks at Blackhall Mill had close connections to the Sword Company, and some connection between these enterprises and Derwentcote certainly cannot be ruled out.

The final outcome of the aborted sale is not recorded, and nor is the outcome of the Chancery suit. However in 1753 Derwentcote was operated by a Mr Hodgson of The Close, Newcastle (Barraclough 1984a, 68–9, 203, quoting Angerstein's diary). Hodgson used the cloth shear trademark formerly used by Hayford for Bertram's steel, and a former apprentice of Bertram's was making German steel at Derwentcote, while his son was steelmaker at Blackhall Mill (which, on my reading of Angerstein (in a translation kindly provided by K Barraclough), was operated by a Mr Hall).

The ownership is considerably clarified in 1767, by the will of Gabriel Hall, 'Sadler', of Newcastle. Hall's assets included a one-fifth share in 'Darwencoat forges and furnaces carried on in partnership by Mr John Hodgson myself and others', and a one-third share in 'Blackhall Mill steel forge and furnace carried on in partnership by John Cookson and myself', along with an oil leather business and the Ouseburn oil mills in Newcastle.

Gabriel Hall, John Hodgson, and the Cookson family were linked in a number of business partnerships (not all related to the iron industry). The origin of their iron and steel businesses can be traced to an iron foundry at Old Trunk Key, Gateshead (taken over by Isaac Cookson and Joseph Button in 1721), and a coke-fuelled blast furnace at Little Clifton, Cumbria (site acquired by Isaac and William Cookson in 1721, and under construction in 1723). In 1729 these partnerships were merged, with the addition of John Williams and Edward Kendall of Stourbridge; Kendall was a leading member of the North Midlands/Lake District ironmaster partnerships (Awty 1957, 98, 111–118), and his presence links the Tyneside partnerships into the web of partnerships that dominated the iron industry nationally. The new partnership was to operate a foundry at Newcastle, as well as the Gateshead and Little Clifton works. A new coke-charged blast furnace at Whitehill, Co Durham (near Chester-le-Street) was added in 1747 (Riden 1987, 35). In 1760, following John Button's death, his share in these works was sold by John Cookson and Gabriel Hall to John Williams (now of Newcastle); the sale notice (*Newcastle Courant*, 26th July 1760) also included a share in an apparently separate foundry in Gateshead, operated by John Hodgson and Co. Old Trunk Key was operated by John Cookson and Co in 1763 (*Newcastle Journal*, 11th June 1763), and Clifton, Whitehill, and Newcastle by Cookson, Williams and Hodgson in 1765 (*Newcastle Journal*, 22nd June 1765); however it appears to have closed by 1766 (GL: CA/2/122–3).

In 1752, John Button leased land next to the foundry, to build a steel furnace; in 1776 the lease was renewed to John Cookson, and in 1792 it was re-let to David Landell and Richard Chambers; it was therefore operated by the same web of partnerships that controlled Derwentcote. (GPL Cotesworth Papers, CA/2/116–122–3, 149, 153, quoted to me by Peter King).

John Cookson, Cookson and Co, Cookson and Hall, Cookson and Hodgson, Cookson and Williams, John Hodgson and Co, Williams and Co were all importing iron (sometimes identified as Swedish) through Newcastle in the mid eighteenth century (TWAS 659/234/235), but these imports cannot be related to any specific works, especially as Cookson was a merchant as well as an ironmaster. Cookson's letterbook for the period 1747–1769 survives (TWAS: 1512/5571), though it is unfortunately unspecific about his works. In 1749 (to Mr John Ross) he already 'consumed a large quantity of orgrounde Iron yearly'; the specification of Öregrund Swedish iron, and (in a later letter) of specific marks, suggests that this iron was used for steelmaking, though this was first stated in 1762. In 1752 he mentions having bought into a forge and foundry the previous spring. Jonathan Kendall (son of Edward) was a correspondent; he remained a partner in Clifton, Whitehill, Gateshead, and Newcastle until 1758, when he sold out.

It is clear therefore that, until Hall's death, Derwentcote was operated as part of a web of partnerships controlling much of the Tyneside iron and steel industries (with the major exception of the Crowley works (Flinn 1962)); these partnerships may have centred on John Cookson, and were linked via Kendall into the wider web of the iron and steel industries nationally. However Derwentcote then passed temporarily out the control of the Cooksons, in a period for which the historical evidence is particularly patchy and at times contradictory.

John Hodgson's partnership in Derwentcote continued until his bankruptcy in 1782, when his moiety of a one-fifth share of the 'Iron and Steel forges' was advertised for sale by his assignees, James Rudman and Richard Chambers (*Newcastle Courant*, 6th July 1782). Chambers was a former associate of Gabriel Hall; in 1760 Hall, Roger Heron, and Richard Chambers had set up a partnership as hardwaremen. In 1778 and 1782, the first Newcastle directories list Landell and Chambers (presumably the successors to this partnership) as hardwaremen and ironmongers, whereas by 1787 they were also oil and steel manufacturers; this state of affairs continued in 1789 and 1790, but the firm had reverted to being hardwaremen in 1795. By 1784, they controlled Derwentcote, writing to Henry Cort (for licence to use his patented puddling process of wrought iron making) 'We have laid in the materials for build[in]g a new [puddling] furnace at Derwent Coat Forge' (Science Museum Library: Weale MSS, 371/3 f 206; I am grateful to Chris Evans for this reference).

By 1794, however, 'Darwincoat' forge was occupied by 'Mr Cookson' (Riden 1987, 35, quoting Boulton and Watt Papers), perhaps on behalf of the Owners of Derwent Coat Forge: on 8th April 1797 the *Newcastle Courant* carried the announcement:

Derwent Coat Forge
The Iron and Steel Business, known under the Firm of the Owners of Derwent Coat Forge, is now carried on under the firm of Isaac Cookson and Co, Steel and Iron Warehouse, Close, Newcastle upon Tyne.

The above evidence should indicate that Derwentcote passed from the hands of John Hodgson, Gabriel Hall's executors, and their partners to those of Landell and Chambers at some date between 1767 and 1784 (perhaps between 1782 and 1784), from Landell and Chambers to a firm dominated by Cookson (perhaps between 1790 and 1794; Landell died in 1793, and Chambers in 1794 (wills in DUDPD)), and to Isaac Cookson and Co in 1797.

However this picture is in part contradicted by the incomplete and equally confusing Land Tax assessments for Medomsley (DRO Q/D/L 26–28; DUDPD Land Tax Box 43/2), which indicate the following proprietors and occupiers:

	Proprietor	**Occupier**
1788	Cookson Esq	For the Warren
	Mr Ruben Richley	For Derwentcoat
	and Co	Forge
1789	Isaac Cookson Esq	John Hepple
	Mr Ruben Richley	For Derwentcoat
	and Co	Forge
1795	Isaac Cookson and Co	Themselves
	for Low Forge	
1797	Isaac Cookson Esq	Mr Ruben Richley
1798	Isaac Cookson	for Low Forge
	Esq and Co	
	ditto for the Warren	Mr Richley

The Warren was the fields immediately north of the forge (Medomsley Tithe Map, 1842, DUDPD); Hepple was resident at the forge in 1754 and 1796, and Richley was a Medomsley man born in 1736 (DRO EP/Me 2, 3). It is possible that Ruben Richley and Co was an otherwise-unrecorded operator, intervening between Landell and Chambers and Cooksons. However it seems more likely that he was merely an agent or manager (perhaps for the 'Owners of Derwent Coat Forge'). The presence of Cookson throughout may indicate that Landell and Chambers had in fact relinquished their interest by 1788.

Management: the nineteenth century

Mentions of Derwentcote in the period 1800–1840 are notably sparse. Land Tax assessments show Reuben Richley as occupier for the Darwen Coat Forge Company in 1802 and 1803. The forge was not named in the 1810–12 assessments, but may have been a plot held by John Armstrong, described as a mill from 1811, but as Armstrong Forge in 1816. From 1825 to 1829, the 'forge lands' were occupied by Isaac Cookson, as tenant. Bailey (1810, 288) describes both Derwentcote and Blackhall Mill as belonging to 'Mr Cookson', and as making wrought iron and blister steel. Newcastle directories for the period 1800–1850 consistently show Isaac Cookson (succeeded by Thomas) as a steelmaker of The Close; some of these entries specify The Close as his office, his works being at Blackhall Mill (1822, 1827, 1828, 1833) or Derwentcote (1847). The 1841 Census Enumerators' Returns list five forgemen and an agent (Henry Jefferson) at Derwentcote, the 1851 Returns adding the information that most of the adults, and children born before 1833, were born in Chopwell (the chapelry containing Blackhall Mill), whereas children born after 1838 were born in Medomsley.

From these lines of evidence, it is likely that Derwentcote was leased by the Cooksons throughout the period, but that it may well have been closed down or mothballed in around 1810 in favour of Blackhall Mill (the forge possibly being converted briefly to a corn mill), being reopened between 1833 and 1838.

Isaac Cookson died in 1831 (NRO ZCK 2/11), leaving a fortune of £387,829. However the bulk of his assets derived from his massive glassmaking interests; his estate included stock-in-trade of wrought iron and steel to the value of £5264, with debts of £6762, and leasehold 'iron and steel tenaments' (sic) valued at £200.

By 1850, Thomas Cookson was in severe financial difficulties, culminating in bankruptcy; his solicitor (John Clayton) advised him to leave the country in 1851 (TWAS 39/21–23). Meetings in March 1850 between Clayton, Thomas Cookson's wife, and their son Charles Edward Cookson, to save 'the Steel concern' (NRO ZCK 2/16) were clearly successful; the W B Lead Company began paying Charles E Cookson and Co for steel in 1850 (NRO 712/B24), and directory entries to this company commence in 1851, the Newcastle office and warehouse moving to South Street by 1853. From 1863 onwards, some references specify Mrs Sarah Turnbull Cookson as the director of C E Cookson and Co.

The Cooksons' control of Derwentcote ended in 1872, the works being carried on by a series of short-lived partnerships linked by the inclusion successively of Thomas and Charles Winter. Thomas Stanley Winter was a commercial traveller in 1859 and 1865 and ran an ironmongery business in 1867–9 (according to Newcastle directories), but by the 1871 Census he lived at Derwentcote House, and was a 58-year-old 'agent'. His partnerships can be traced from the tradesmen's accounts of the Beaumont company (NRO 712/B6, B26, and B27); they made quarterly payments to C E Cookson and Co until March 1872, to Scott and Winter from June 1872 to September 1873, to Ramsey Winter and Co from March 1874 to September 1875 (the dissolution of this partnership was announced in the *London Gazette* for December 7th 1875 (Guildhall Library, pers comm)), and to Charles H Winter from December 1875 to June 1878. Steel purchases then became erratic (due to the decline of the lead industry), but there were purchases from the Derwent Cote Steel Co Ltd in 1880 and 1882; the 1881 Census shows that a workforce was still present at the forge (*see below*). Some Newcastle directories show the presence of this company at South Street from 1884 (when George C Barker was managing director) to 1890, and Charles Winter is listed as proprietor of Derwent Cote steel works (with no mention of South Street) in 1894. The Beaumont company made a single payment for steel to Charles Winter in December 1895. Land Tax assessments (DUDPD Land Tax Box 43/20 onwards) confirm the above picture for the 1870s; they show the Derwent Cote Steel Co as occupiers from 1879 to 1891, and C H Winter from 1891 to 1896. The 1891 Census Enumerators' Returns confirm that Charles Winter was living at Derwent Cote, with his wife and six daughters; the other residents were listed as labourers or clearly non-relevant occupations, though one forge labourer (Stephen Hanson) of Coltparkhaugh in Blackhall Mill may possibly have worked at Derwentcote.

Taking this evidence together, it is clear that Derwentcote remained in the occupation of the Derwent Cote Steel Co until 1891, though production did not necessarily continue throughout this period. It is very doubtful whether Charles Winter actually produced any steel after 1891; the one recorded sale could well have been old stock.

The workforce

The earliest specific mention of the workforce comes from the 1740s Chancery suit: between 1746 and 1748 George Blenkinsopp tried to eject Joshua Cotton, Joshua Copley, George Beavins, Thomas Elteringham the elder and younger, and John Ward from the disputed leasehold (PRO: C 11/381/113; see above). A George Bevins and John Lavender were brought over from Ireland in 1713 to act as forgemen for the Backbarrow Iron Company; the presence of a John Lavender in the Medomsley Parish Registers in 1719 and 1720 strengthens suspicions that this is the same man (or his father), and that Bevins and Lavender were among the first forgemen recruited for his new forge by Ralph Reed (J Swabey, pers comm).

Remaining information on the eighteenth century workforce at Derwentcote comes from the Medomsley and Ryton parish registers; it is limited in quality, since abode and occupation are rarely stated in the Medomsley registers (especially earlier in the century), and the former is sometimes ambiguous (since wording such as 'the forge' or 'Low Forge' could refer to either Derwentcote or Blackhall Mill, unless Medomsley is specified).

There are no proven references to Derwentcote employees before the 1750s, though Eltringhams, Gelleys, Richleys, and Wards (all families later working at Derwentcote) were living in the chapelry. Bartrams, Cawthorns, Eltringhams, Shaws, Vintins, and Wards can however all be identified as working at Blackhall Mill in the earlier eighteenth century (and a number of Eltringhams lived at Swalwell, perhaps working at Crowley's works). From the 1750s, residents (though not necessarily workforce) can be identified; Caleb Atkinson, Stephen and Eleanor Eltringham, Richard Gelley, and John Hepple are mentioned. Richard Gelley was a forgeman at Derwent Coat in the 1780s. Many of these families recur in the 1790s, with the additions of William and Catherine Cawthorn (probably of the Blackhall Mill family) and Robert Summerson. William Gelley was still at Derwentcote in 1809. Both Blackhall Mill and Derwentcote had a forgeman called Stephen Eltringham, though not necessarily the same man. In 1825, Joseph Hepple was making German steel for Cooksons at Blackhall Mill (Billington 1825).

From 1841 onwards, the Census Enumerators' Returns for Medomsley give more detailed information. It is clear that some residents, both at Derwentcote and at the forge itself, had unrelated occupations; conversely forge workers living away from the site may not have been identified.

In 1841, Henry Jefferson was the Agent, and there were five forgemen: Joshua Atkinson, George Elteringham (whose son Thomas, aged 20, had no listed occupation), two Stephen Elteringhams and Spencer Smith. Robert Elteringham (mason; son of the older Stephen), Richard Gelley and Michael Cawthorn (carpenters), and Michael Mitchelson (charcoal burner) may also have worked at least in part for the iron and steel works. Three forgemen living at Blackhall Mill may still have been operating this works, or have walked to work at Derwentcote.

In 1851, Henry Jefferson was still Agent. Stephen Eltringhan and Joseph Gelley were forgemen, but Spencer Smith and his sons William and Robert were specified as steel forgemen. Thomas Eltringham was a steel converter (the operator of a cementation furnace), and George Eltringham (probably the 1841 'forgeman') was a retired steel converter. Anthony Fewster was the forge carpenter, and Thomas Pigg was a quarryman living at the forge.

In 1861 there was no resident Agent. Joseph Gelley and George Harrison were iron forgemen, and Stephen Eltringham and Ralph Elliot were steel forgemen. Thomas Eltringham was still steel converter, and also roller, and Joseph Brown (born at Newburn) was steel melter (the operator of a crucible melting shop). James Scullin was a quarryman, lodging with Robert Eltringham (now a builder).

By 1871, Thomas Winter was living at Derwentcote House. There were no iron forgemen, the steel forgemen being George Harrison, his son Edward, and Matthew Wright. Thomas Eltringham was still the steel 'maker', and Robert Broadsword was forge carpenter. No steel melter was identified.

By 1881, there had been a complete turnover of the workforce. The 'steel manager' was John Long, from Wortley near Sheffield. The forgemen were Benjamin Storey and his sons James and Matthew (all from Winlaton). Henry Godley was steel melter; he was from Nottinghamshire, but his children were born in Sheffield. Three labourers were probably not connected to the works. In addition, George Barker, a retired Newcastle ironmonger, lived at Derwentcote; he was probably the managing director of the Derwent Cote Steel Company mentioned in 1884 (see above). In 1891, as already mentioned, Charles Winter lived at Derwentcote as a 'steel manufacturer', but had no identifiable workforce.

It therefore appears that the workforces of Derwentcote and Blackhall Mill were very closely linked, and showed great family continuity from the early eighteenth century until the mid nineteenth; all the families in the 1841 workforce had eighteenth century antecedents. These familes gradually disappeared between 1841 and 1871, being progressively replaced by incomers (though all from within the Tyneside region). Between 1871 and 1881, however, there was a complete turnover; the manager and melter were apparently brought in from Sheffield, while the forgemen came from Winlaton.

The movement of workmen, and perhaps of skills, from Derwentcote is inevitably not amenable to analysis from local parish registers; a full analysis of the inter-relationships and movements of workforces between the steelworks of the Derwent Valley area, and of migrations between the works and the Sheffield area, might yield interesting information. It may be noted that the manufacture of 'shear steel' (Bertram's introduction to England) was apparently introduced to Sheffield by a Thomas Eltringham from Blackhall Mill in 1767 (Barraclough 1984a, 66).

Technological development and trade

The general technology of iron and steel making has been described in Chapter 2. It should be noted again that the Derwentcote works contained both a cementation furnace and a forge; the former was specifically and exclusively a steel producer, whereas the latter could and (from documentary evidence) did produce wrought iron as well as steel, and the two aspects should be considered separately. Crucible steel production is considered as part of the forge activities.

The cementation furnace produced blister steel from Swedish bar (wrought) iron. The process as used at Derwentcote was described by Angerstein in 1753 (Barraclough 1984a, 68, and pers. comm; it should be noted that the illustration reproduced as fig 13 (68) in Barraclough 1984a is now believed to be of Blackhall Mill not Derwentcote). The 'pots' [cementation chests] held 10–11 tons of Öregrund iron, packed round with charcoal; each heating took six to seven days, using 40s worth of coal, some of which could be re-used [this must therefore mean charcoal], and was followed by 12–13 days cooling. There were only two workers at the furnace; the senior steel forgeman was a former apprentice of Bertrams's, and made a few tons annually of 'so-called German [presumably shear] steel'. Production of blister steel was just over 100 tons; the limiting factor was economical supplies of Öregrund iron. Angerstein claimed that most of the blister steel was merely drawn down into bars, and dispatched to the East India Company; however examination of the East India Company's accounts (which survive complete for this period) shows absolutely no evidence of steel purchases from any of the partnerships associated with Derwentcote, and it seems likely that Angerstein was misinformed.

There is however a hint of sales to Sheffield at a slightly later date. In 1766, the creditors of Samuel Bishop (a Sheffield shearsmith) included John Hodgson and Co, for two purchases (each of two faggots of steel priced at £5-4-0) in December 1765 and January 1766; these were presumably the two bundles of 'German steel' whose freight from Hull to Sheffield appears in Bishop's debts (SRO TC 1043–5). Since Derwentcote is the only steelworks known to have been operated by Hodgson at this date (Hall and Cookson operating Blackhall Mill), this may well have been Derwentcote production.

The furnace was built before 1742, perhaps in or just after 1733 (*Ch 4*), and the presence of Thomas Eltringham as steel converter or maker in the 1851–1871 censuses shows that it was still in production. Evidence for use of the furnace after 1871 is ambiguous. The absence of a converter in 1881 suggests that the cementation furnace had been abandoned by this date, and this suggestion is strengthened by the absence of definite cementation products in the sales to W B Lead after 1872 (*see below*). However a billhead of 1888 (Beamish Museum, Wears of West Boat Papers) still refers to the Derwent Cote Steel Company as 'Steel Converters and Refiners, Manufacturers of all kinds of Cast, Shear, Blister, Spring, and Mining Steel, Cast Steel Hammers, Crucible Steel Castings, Iron and Steel Forgings etc', suggesting that (at the least) use of the steel furnace was still envisaged when the Derwent Cote Steel Company was set up in 1879.

The forge was in existence by 1719 (*see above*); presumably it was a normal water-powered finery/chafery forge, converting pig to wrought iron. By 1736 it was producing 125 tons of (bar) iron per annum (Hulme 1929, 25). The list of buildings in 1748 has already been quoted (*Ch 4*). In 1753, according to Angerstein, it had a finery (using charcoal collected from a wide area, at a cost of 4s 6d per 10-bushel horseload), a chafery (using pitcoal), and a hammer; it produced 150 tons of bar iron per annum from English and American pig iron, this iron being used locally for making firegrates for steam engines, axles for coal waggons, and other purposes not requiring high quality iron.

In the second half of the eighteenth century, considerable efforts were made to develop a more rapid and economical alternative to the finery/chafery process, for converting pig to wrought iron (described in Mott 1983). These can be summarised as the development first of the 'stamping and potting' process, superceded in the 1780s by the puddling and rolling process perfected by Henry Cort, which used raw coal in a reverberatory furnace. This latter process formed the basis for the explosive rise of the British iron industry, from the end of the eighteenth century until the development of Bessemer and other mild-steel making processes in the mid nineteenth century.

At Derwentcote, the forge was a very early user of Cort's process (Science Museum library, Weale MS 371/3). In June 1784, Landell and Chambers had 'taken a privilege' to make iron by Cort's process (f 205); in November they wrote to Cort requesting a plan for a [puddling] furnace to be erected by their own workmen at Derwentcote forge (f 206). In 1788, however, John Cooke (of Kilnhurst Forge, near Doncaster) informed Cort that at Derwentcote

> the practice is to make Iron in the Test furnace, but not to ball it therein, but to take it out by small pieces and stamp the cinder out, and when cold to pile the thin pieces on thin stones and heat them in piles of $\frac{1}{2}$ cwt each in another Air furnace. (f 202)

'Stamping' in this instance presumably refers to hammering the plates to drive out slag, rather than to breaking up the plates into fragments (the meaning of the term in the stamping and potting process); reheating piles of plates in an air (reverberatory) furnace was a common modification of stamping and potting, patented by Wright and Jesson in 1784 (Mott ed. Singer 1983, 13), but the use of thin stones is hard to understand, and Cooke may have been misinformed on this point.

By 1794, the forge contained two fineries, a chafery, and a balling furnace (Riden 1987, 35, quoting Boulton and Watt Papers). This seems to be a reversion to more orthodox technology, suggesting that the excursion into puddling may not have been successful. In 1810 Bailey (1810, 288) states that at Derwentcote and Blackhall Mill,

> 'they make malleable iron from old cast metal, and fuse old iron in a furnace, to make into sock moulds, shovels, boiler plates, bar iron etc; they also make cementation steel by cementation with charcoal, and work it over afterwards four or five times with hammers'

though it is not clear that all these activities were undertaken at both sites.

As discussed above, there is some evidence that Derwentcote (forge and furnace) may have been closed down from *c* 1810 to the mid 1830s; if so its re-opening would provide a logical context for modernisation, but positive evidence is lacking. Sales to the WB Lead Company between 1844 and 1850 (NRO 712/B24) include cast iron (presumably from the Newcastle foundry), bar and sheet iron, and blister and German steel (all presumably from Derwentcote). In 1851, Census evidence confirms that both iron and steel were being forged, by separate workmen and therefore probably in separate workshops at the forge. Iron forgemen were still present in 1861, but Cooksons disappeared from the list of 'iron manufacturers' in the Newcastle Directories in the late 1850s, and by 1871 there is no evidence that iron (as opposed to steel) working still continued.

In contrast, the first evidence for cast steel manufacture occurs in the 1861 Census, where a steel 'melter' is identified; Spencer (1864, 125) lists Derwentcote as having six 'melting furnaces', and makes it clear that crucible steelmaking in the Northeast still used a blister steel charge, as opposed to the charges of mixed Swedish wrought and cast iron, which by this date were tending to replace blister steel for this purpose in the Sheffield area (Barraclough 1984b, 63–71). The absence of a steel melter in 1851, and of cast steel sales to W B Lead in the period 1844–50, suggest that crucible melting was introduced to Derwentcote in the 1850s. The addition of 'steel roller' to George Eltringham's title in 1861 may point to the construction of a rolling mill (presumably water-powered) in the same decade; from 1853 onwards Directory entries sometimes specify C E Cookson and Co as steel spring (and in some entries file) manufacturers, and Spencer

(1864, 122) states that spring steel was produced by rolling of blister bars. It therefore appears that C E Cookson modernised the works by the addition of rolling and crucible melting, with a partial specialisation in spring manufacture, while the wrought iron business declined.

As already noted, evidence for cementation steelmaking and its products dies out in the early 1870s. Detailed records of sales to W B Lead's Allen Smeltmill, from 1869 onwards (NRO 712/B25, B26) include German steel and shear steel in 1872, but no later identifiable cementation products. Between 1869 and 1873, 36 steel ladles of various sizes and perforations were supplied, one batch being specified as cast steel; from 1875 to 1877 a further 19 steel ladles were supplied, and one 'wrought' ladle. Twelve bars of mild steel were also supplied; these had presumably been bought in by Charles Winter. Steel was also supplied to other divisions of W B Lead (NRO 712/B27), including 'steel castings' in 1874, and 'jumper steel' (for hand-drilling chisels) in 1880 and 1882.

It is therefore likely that the cementation furnace was closed in the early 1870s, while crucible steel manufacture continued (probably using charges of mixed cast and wrought iron, rather than of blister steel), the crucible steel being made into both castings and forgings.

5

THE STANDING STRUCTURE

The standing structures at the furnace site consist of four structural units:-

1   The furnace itself, a heavily-built and heavily-buttressed rectangular structure at ground level developing into an elongated cone at higher level, with considerable internal and external complications.

2   The Northern Building, a rectangular structure butted onto the north face of the furnace.

3   The Southern Building, a longer rectangular structure butted onto the south face of the furnace.

4   The Eastern Building, a small lean-to building occupying the angle between the Southern Building and the southern buttress on the eastern face of the furnace.

The first three units are symmetrical (with slight deviations noted below) around a single north-south axis, which also formed the ridge-line of the roofs of the Northern and Southern Buildings. At the time of survey, the furnace stood virtually intact except for damage to some of the buttresses and the top of the cone; the three ancillary buildings were roofless, but most of their walls stood to eaves/gable height (Plates 5, 6). These buildings have since been reroofed, and the furnace conserved.

The normal method of graphic recording was by rectified photography following cleaning of the walls, as a base for stone-by-stone and interpretative drawings at 1:20, which were checked and augmented on site. Computer photogrammetry from a hydraulic lift was used to record the outer faces of the furnace cone, and other areas inaccessible to rectified photography were drawn by hand. Written recording was undertaking using a context system (contexts 500–800 and 1200–1232); these context numbers are used in the following text only where they assist in clear referencing. The interior spaces of the furnace, and other information observed during consolidation work after the main recording programme, were recorded in narrative text form without the use of context numbers.

Mortars. Most of the mortars and other bedding materials could be divided by eye into five consistent Types (I–V):

Type I. Homogenous soft yellow-brown (Munsell 10YR 5/4–5/5) fine sand and silt, with rare pebbles and coal flecks and slight clay content.

Type II. Hard grey (10YR 5/1–6/1) cement-based mortar with sparse but large lime flecks, and some coal and brick/tile fragments.

Type III. Rather variable pale brown to pale pink (10 YR 8/3–5YR 8/3) lime-sand mortar with pieces of coal ash and slag; consistency varied from hard via crumbly to soft and sandy.

Type IV. Grey (N 4.5/0) greasy clay with lime and coal flecks.

Type V. Light grey to brown (10YR 7/2–5/3) lime-sand mortar (often decayed to sand with lime flecks), with coal and brick/tile fragments, and occasional pantile pieces.

The furnace itself was bedded and pointed in Type III mortar, with alterations and repointings in Type IV, whereas the three ancillary buildings were bedded in Type I, and pointed with Type V (though this only survived in patches, having largely weathered off on the more exposed faces). Type II was associated with late alterations and repairs to various parts of the structure.

The furnace

The furnace formed the spatial and functional core of the buildings, and was structurally the earliest, being abutted by all three ancillary buildings (Figs 2–7).

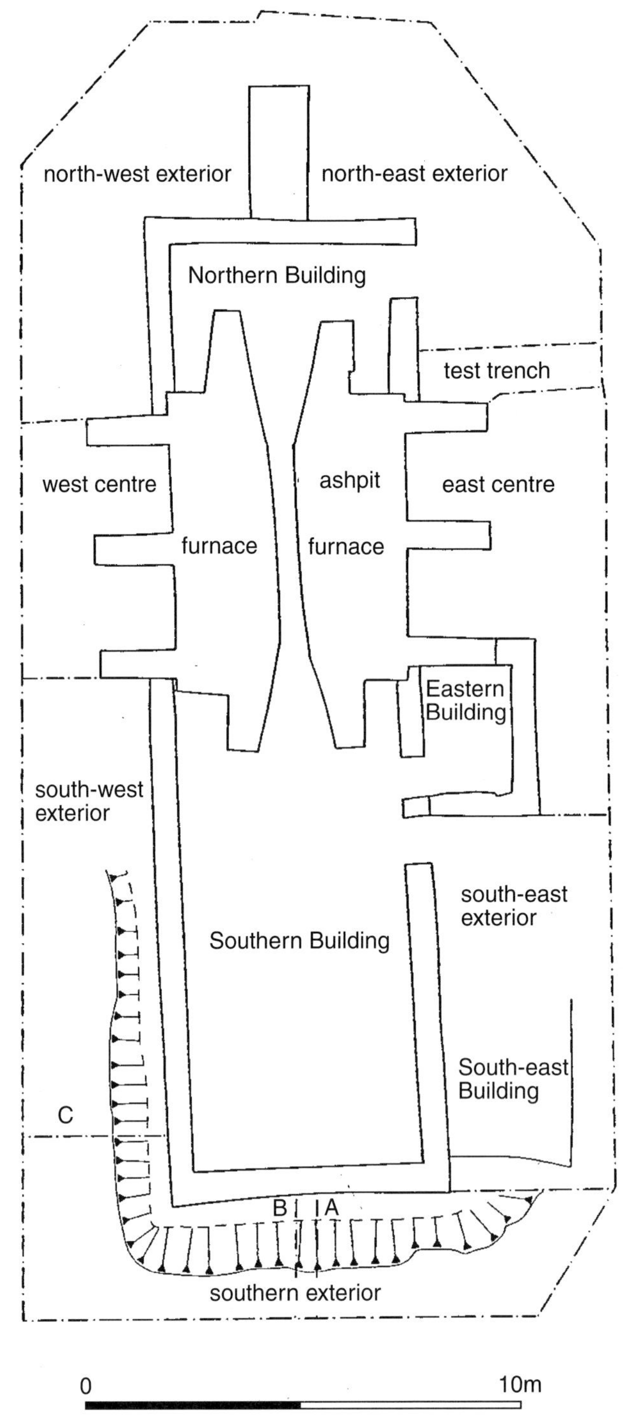

Fig 3   Overall plan of excavation, showing area locations

Externally, the form of the structure fell into two parts. At ground level, it was rectangular and vertical-walled, measuring 6.8 m north-south by 5.6 m east-west. Three buttresses projected from each of the long sides, and two from each end. The latter were trapezoidal in plan, their splayed inner faces forming the sides of a central arch in each end face, giving access to the firegrate and ashpit along the axis of the furnace (*see below*). Above these arches the wall faces were continuous, but were bowed outwards in plan. This rectangular part of the structure was divided, on all four faces, by three slight horizontal plinths; the first of these was just above the surrounding floor and ground levels, the second was *c* 1.8 m above the first, and the third was 0.9 m above the second.

Above this level, the rectangular plan of the structure was carried into an oval cone, truncated at the top. On the north and south ends, the junction was formed by platforms and plinths of varying width, the base of the cone resting on the (bowed) top of the rectangle. On the east and west sides, the junction was formed by a curved arris between the lower part of the cone and segments of vertical walling carried up above third plinth level.

The internal form consisted of three superimposed chambers. The lowest of these (the ashpit below ground level, and the firegrate close to external floor level) formed an elongated dumb-bell in plan, along the north-south axis of the furnace. This opened upwards into a vaulted rectangular chamber (the chest compartment) measuring 5.2 x 3.2 m. This was separated by the vault (at approximately the level of the third external plinth) from the oval interior of the cone.

Apart from technical difficulties of recording, interpretation was complicated by four problems. Firstly, the condition of the external faces varied: while the upper parts and the buttresses were composed of hard stone with clear weathered-out joints, much of the remaining walling consisted of soft friable stone set in hard mortar with a surface crust that could not be removed without damage to the stone (this difference was probably due to the prolonged heating to which these areas of the structure will have been subjected during the use of the furnace). Secondly, the Type III mortar showed variation in its inclusions as well as in its surviving state; the variations could not be correlated with any plausible phasing, and were probably merely batch differences. Thirdly, the structure had undergone considerable movement, shown by wide sub-vertical cracks and by features interpreted as horizontal shearing; these inevitably tended to follow pre-existing joints of various types. Much of this movement had occurred during the working life of the furnace (as shown by heat effects and/or patching in the cracks), and was presumably the result of thermal expansion and contraction, though later 'spreading' (perhaps due to mining subsidence) had also occurred. Fourthly, some of the structural junctions were themselves rather odd in character even before being affected by secondary movement (perhaps because they were designed to allow for such movement).

The structural evidence described below was therefore complex and hard to interpret; it should be stated that, while limited repairs and alterations have been detected, hypotheses involving major rebuilding have been explicitly examined and rejected.

The exterior

The east and west faces were very similar in form, consisting of vertical walls divided horizontally by the three plinths, and ending upwards at a clear curving arris between the straight walling and the convex walling of the cone (the two being structurally continuous). Each wall was supported by three evenly-spaced buttresses, the southern and northern corners of the face extending 0.2 – 0.3 m beyond the outer sides of the respective buttresses. Each wall also contained two sub-square openings (the 'loading holes') just below second plinth level on each side of the central buttress, giving access to the internal chest compartment. The walls were composed of coursed rectangular sandstone rubble bedded in Type III mortar, the heights of the courses tending to increase upwards. The corners were composed of rather larger blocks, but without distinct quoining. Dressed faces survived in the most sheltered areas.

On the west face (Figs 4, 5, Plate 7), the plinths were respectively 50 mm, 80 mm, and 110 mm wide; above the third plinth, the segment of straight walling was battered back by 0.12 m in its maximum height of 1.4 m, and was featureless. Cracks in the wall were present, but were quite limited, and were all packed with mortar.

The southern loading hole was blocked. The opening itself was irregular, measuring 0.44 x 0.38 m, having probably worn back from a rectangle of 0.36 x 0.24 m; the sides were burnt and friable. The sides and top were formed of inserted masonry; some of this could be interpreted as blocking of an earlier opening centred slightly further north, but the insertions are more likely to have been mere replacements of the lining of the existing opening. The blocking was composed of brick, firebrick (some vitrified), and pantile, set in brown lime mortar with coal and sand inclusions.

The northern loading hole was open, measuring 0.44 x 0.36 m in its final form. The sides were formed by single thicknesses of inserted firebrick walling bedded in grey silty clay with sandy lime patches, and the top was lintelled with firebrick slabs measuring 530 x 470 x 80 mm. The top and north side of the firebrick abutted cut-back original walling, whereas the south side abutted a stopped end in the original wall, only 0.18 m from the central buttress. This may well have been the south side of an original loading hole closer to the buttress.

The three buttresses were each 0.6 m wide, their west ends being inclined at 28° to the vertical. They had originally extended to flat tops c 0.3 m long at third plinth level. Their sides were of similar build to the furnace wall, whereas the west ends were of larger more regular blocks, some extending the full length

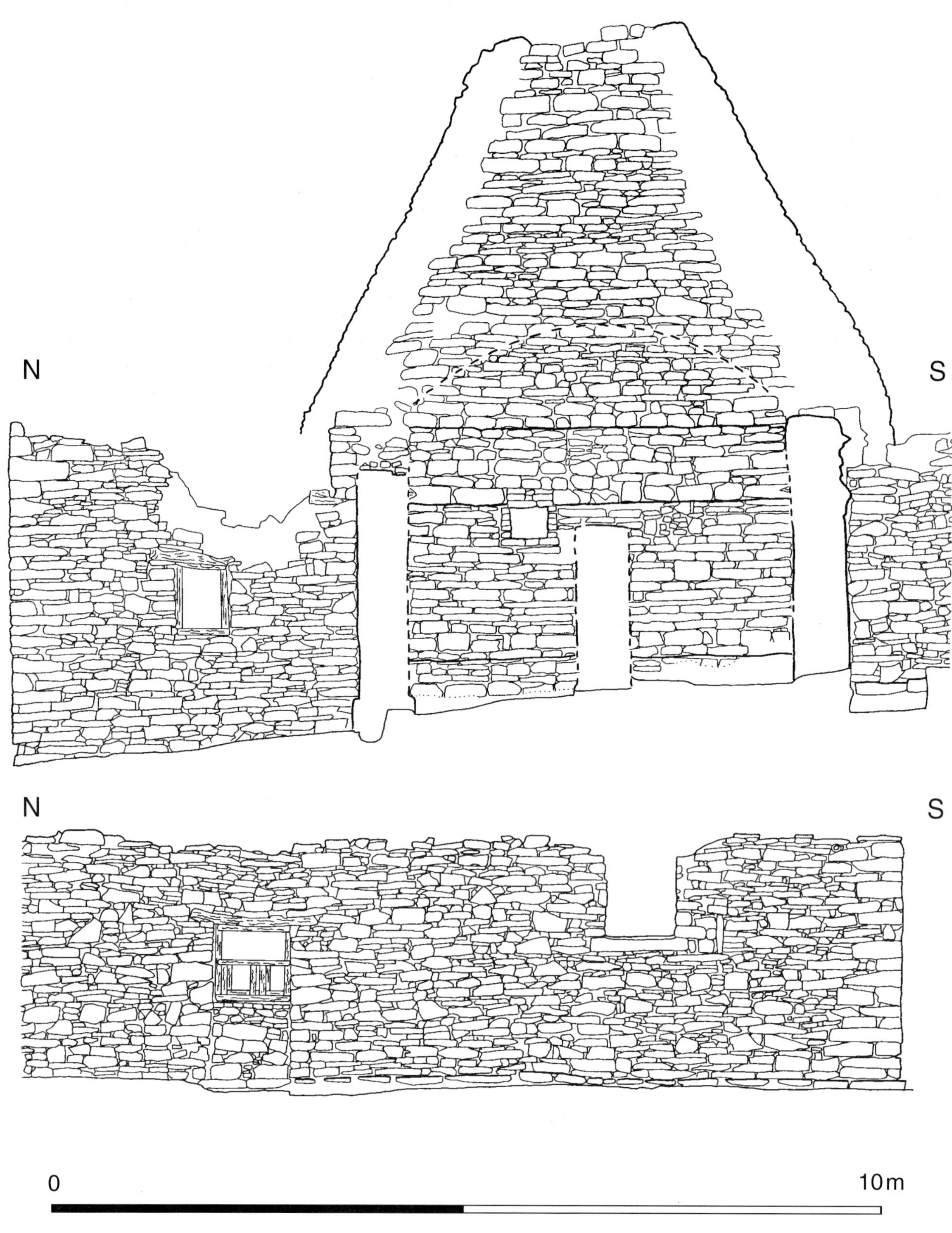

Fig 4   Detailed elevation of west face of buildings; (see Fig 6 for key)

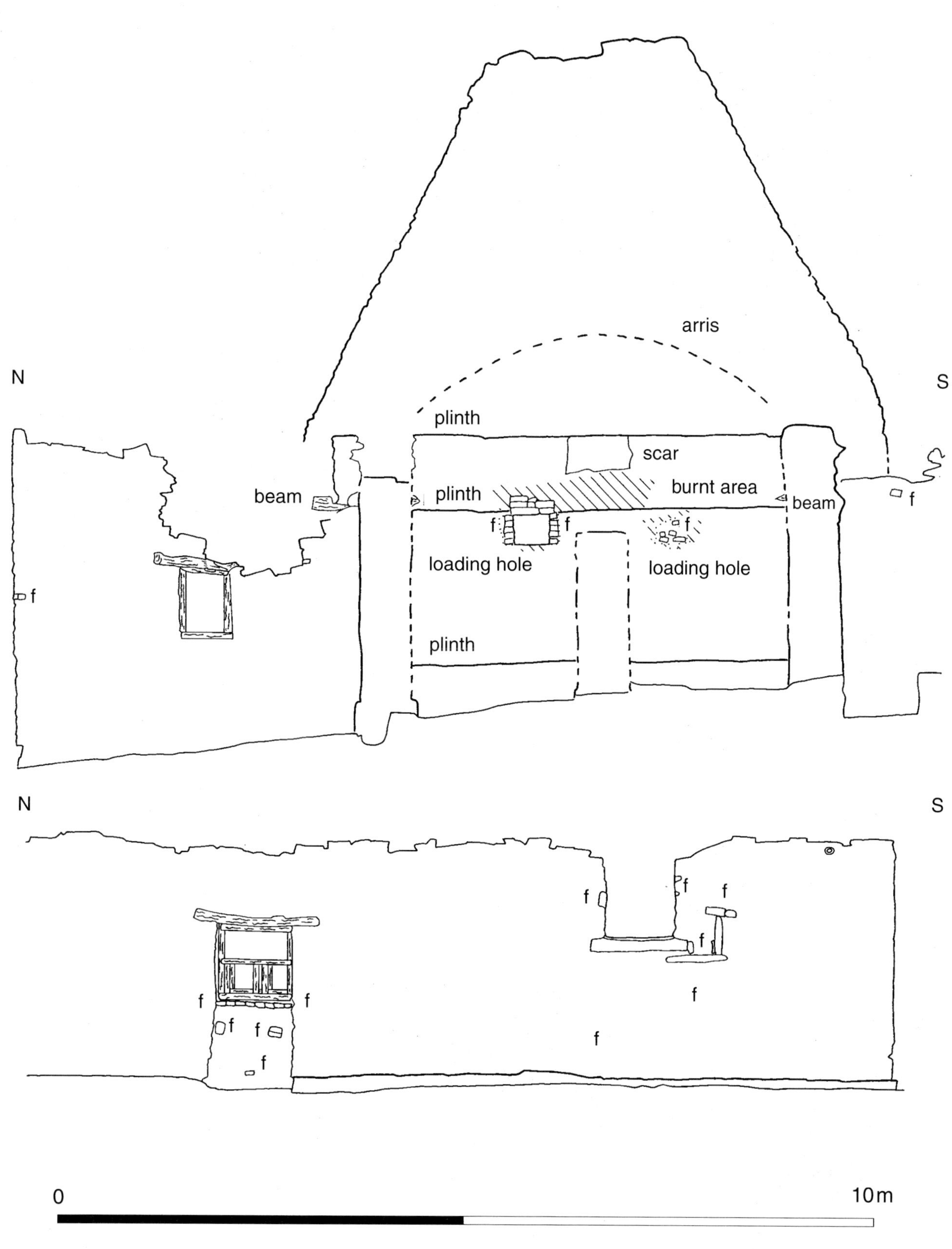

Fig 5   *Elevation of west face of buildings: interpretation*

of the face. The junctions with the furnace wall were at first sight inconsistent, since in places the buttresses abutted the furnace, while in others the furnace wall abutted the face of a buttress, and a few lengths of fully keyed-in joint were also present. This is interpreted as deliberate block-keying, perhaps to allow for differential thermal movement of the furnace. The central buttress was broken off upwards *c* 1.9 m from original ground level (close to the local ground surface before excavation), while the northern and southern buttresses were virtually intact. The west walls of the Northern and Southern Buildings abutted the outer sides of the respective buttresses, the northern buttress containing at least one projecting tusking stone to receive the wall.

The two intact buttresses both contained the rotted remnants of a beam, set horizontally along the face of the furnace wall at second plinth level, within sockets through the buttresses (Plates 8, 9). The beam was 150 – 200 mm square, with chamfered upper arrisses. It had rotted off where it protruded from the buttresses, except to the north of the northern buttress, where it continued for 0.60 m within the core of the Northern Building wall (the end was rotted, but appeared to be original). This projecting part contained a sawn-in rebate 260 mm long by 90 mm deep in its upper surface, partly obstructed by the corner of the furnace; this may have been the remains of a halved joint with an east-west beam along the north face of the furnace. A saw cut at the rotted south end south of the southern buttress may have been the remains of a similar rebate. In this buttress, much of the socket around the beam had been replaced by a patch of inserted sandstone masonry with firebrick fragments, bedded in a variant of Type V mortar. In the northern buttress, however, the socket was 0.45 m high, and was largely infilled with burnt brick and sandstone rubble (some of the latter being vitrified), behind the original sandstone facing; this material appeared to be an original packing rather than a later repair.

Two sockets in the inclined west face of the northern buttress are described below (*Ch 5*) in conjunction with the Northern Building, since they are interpreted as relating to a lean-to structure against this.

The central buttress did not survive to the level of the beam, but a gap in the scar on the face of the furnace implied that a similar socket had been present. This part of the wall face was consistently reddened by fire, the burnt area being up to 0.5 m high and extending for 2 m along the face immediately above the second plinth; it also extended round the top of the northern loading hole. The reddened area can best be explained as originating from the burning of a horizontal beam along the face of the furnace, of

which the lengths embedded in the buttresses were the remnants.

The east face (Fig 6 (a), Plate 10) was very similar to the west. The plinths were respectively 80 mm, 100 – 140 mm, and 80 mm wide. The wall face contained several sub-vertical cracks, infilled with decayed mortar and earth, especially above the two loading openings; the sides of one of these cracks were reddened by heat. The (block-keyed) joints between the northern and central buttresses and the main wall had pulled open to form wide near-vertical cracks, partially packed with decayed mortar and brickbats. At one point, the main wall had moved 0.1 m outwards relative to the buttress.

The southern loading hole (Plate 11) was open, and was 250 mm square in its final form; its structural history was hard to interpret due to the very encrusted state of the wall. The masonry of its north side consisted of burnt sandstone bedded in grey sandy mortar with coal inclusions; it abutted a stopped end in the primary wall, and may have infilled a blocked original loading hole 200 mm square, and only 140 mm from the central buttress. The south side and base were formed by secondary masonry of sandstone rubble set in soft pink-white sandy mortar with lime and coal inclusions, separated from the primary masonry by wide mortar-filled joints; this is more likely to have been a replacement of the lining of the opening than the blocking of a separate feature.

The northern loading hole was also open, except for a sandstone slab recently inserted to prevent collapse of its broken and unstable top (which restricted close examination). It had been enlarged to *c* 0.46 x 0.4 m; the south side and top had been cut (or broken) back during use, and firebrick slab lintels measuring 600 x 470 x 80 mm inserted, bedded in grey-brown clayey mortar, and pointed with hard, white mortar.

The buttresses had been very similar to those on the west face. The southern buttress was intact except for some patching to its top, and had been incorporated into the north wall of the Eastern Building (*see below*); its north face had been partially repointed with Type V mortar. The central buttress was broken off only 0.5 – 1.1 m above the first plinth (the very weathered scar on the furnace above this level suggesting that the damage was not recent); it incorporated burnt masonry, and irregularities in its faces had been packed with pantile, firebrick, and mortar (including Type V). The northern buttress was broken off at second plinth level. Its north face had been partially repointed with Type V mortar, and was abutted by the Northern Building; a gap or socket on the corner of the furnace beside the buttress may have been a 'reverse tusking' for the Northern Building. The

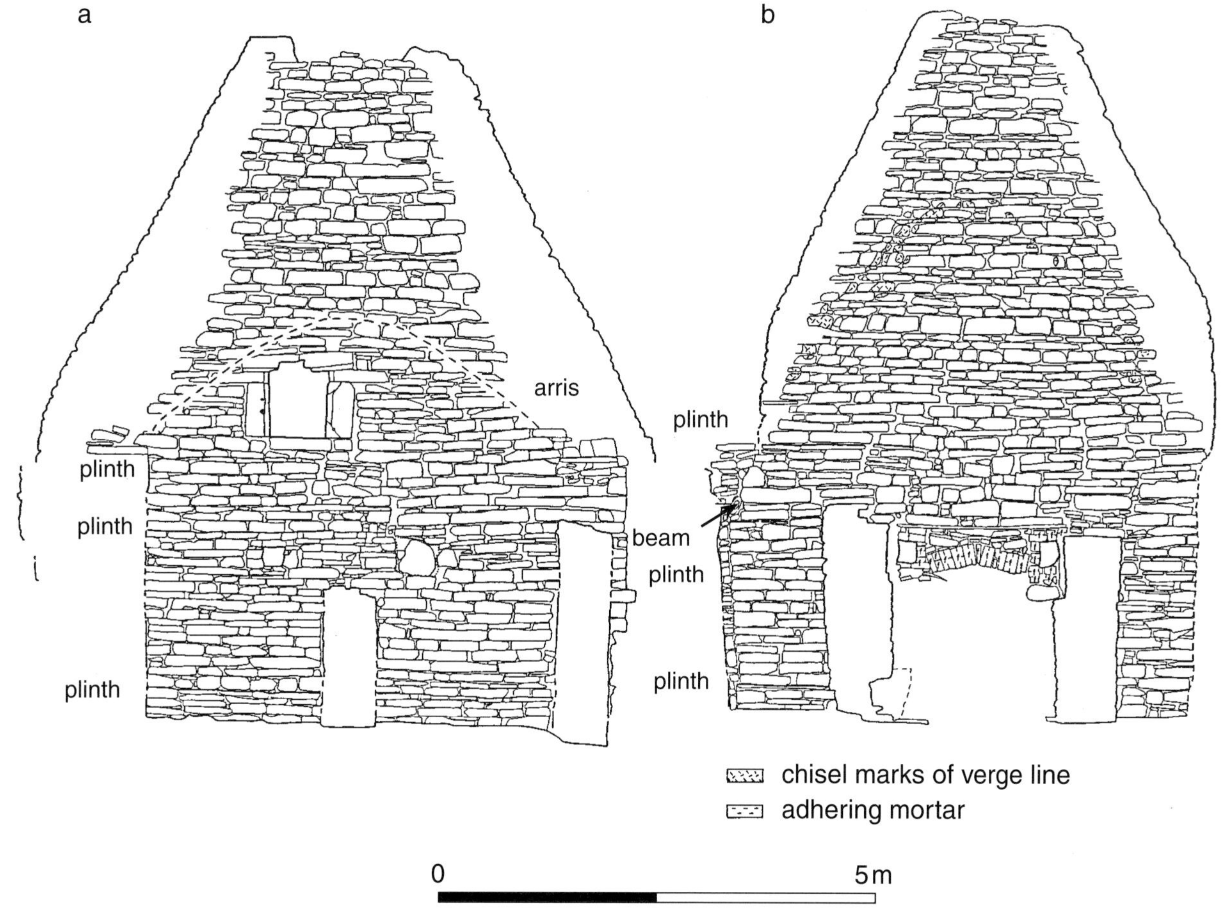

Fig 6   Elevations of Furnace: (a) east face, and (b) south face

northern and southern buttresses rested on projecting rubblestone footings 0.3 m below the first plinth (the equivalent level on the central buttress was not exposed).

The northern and southern buttresses both contained sockets running through them at second plinth level. The socket in the southern buttress (Plate 12) was 0.25 m wide and 0.28 m high, and was infilled with sandstone masonry bedded in grey-brown sandy mortar with shale fragments, and pointed with Type V mortar. Only the lower part of the socket in the northern buttress survived; this was 0.27 m wide and at least 0.4 m high. It had been infilled with a faced blocking of brickbats (including firebrick) set in Type V mortar.

The wall face was burnt red for a length of 1m, and height of 0.3 m, at the level of the sockets, the area of burning connecting to that round the northern loading hole and a large crack above it. As on the west face, this distribution is best explained as the result of burning of a beam set in the sockets.

Above the third plinth, the segment of straight face was slightly battered (by 0.10 m in in its maximum height of 1.5 m), and was separated by a curved arris from the near-circular cone. It contained a single feature (not matched on the west side), to the south of centre: an opening into the interior of the cone. This was 0.80 m high and 0.73 m wide, with quoins and lintel of ashlar, with rebates 60 mm wide along the edges of the opening. The sill was 0.2 m above the third plinth. The quoins contained two small sockets, 25 mm square, on each side just outside the rebate, one of which retained an oval loop of iron set in a lead plug filling the socket. This feature formed the only access to the cone interior; remains of an iron-plated door that may have fitted it were found in the excavated stratigraphy below (*Ch 6, 7*).

The north and south ends of the furnace had also been closely similar to each other, though they had been altered in different ways. They were each dominated by a central lintelled opening (the 'firegrate arch'), over the ashpit and splaying out from the mouth of the firegrate. The sides of this were continuous with the inner faces of two wide buttresses of trapezoidal plan (their outer sides being at right angles to the main walls of the furnace), extending to second plinth level. These main walls, from the buttresses to the corners of the furnace, were rectilinear at ground level; however they were stepped slightly out upwards, continuing as a single gently-bowed wall above the top of the arch and buttresses. The masonry and bedding were similar to those of the east and west faces.

The south face (Fig 6 (b), Plate 13) was exposed to the floor level of the Southern Building, 0.28 m below the first plinth; its top was formed by the third plinth, on which the strongly-curved cone structure rested.

The east end of the wall was straight and vertical at ground-level, the western part being cantilevered out upwards to a bowed line; the first and second plinths were absent. The extreme east end was obscured by the east wall of the Southern Building, which abutted the corner of the furnace. The west end differed in that all three plinths were present, and that the corner of the furnace was exposed (the west wall of the Southern Building abutting the buttress on the west face of the furnace, with a narrow strip of masonry infill between its inner face and the corner of the furnace). Its course ran slightly south of east, at an angle of 95° to the west face, and it did not align with the east end of the face. The wall was cantilevered out upwards, onto a gently-curved line, but it was not clear whether this was deliberate or the result of an outward movement due to thermal stress.

Above the second plinth, the wall face was continuous for the whole length of the side, with a uniform convex curve. The central part, above the lintel of the arch and a repair above this, was separated by wide cracks from the remainder and may have been a rebuild, blocking the original shaft behind the lintel (*see below*). Some of the cracks showed evidence of being filled at least twice, with different mortars.

At second plinth level, the beam projecting from the socket of the southern buttress of the west face (*Ch 5*) was exposed at the corner of the furnace. From this point east to the top of the western buttress on the south face, a horizontal cut-in rebate extended along the south face, its base being level with the sawn 'halving' of the beam.

The western buttress was 1.5 m long, and 2.0 m high from floor level; its width increased from 0.7m at the south end, to 1.1 m on the line of the main furnace wall, due to the (slightly convex) splay of its east side, at *c* 15° to the axis of the furnace. The south face was formed of the same horizontally-coursed masonry as the remainder, and was inclined at 10–20° to the vertical, the angle increasing upwards. The whole buttress was skewed slightly to the west, the west and south sides being displaced at an irregular line 0.2–0.6 m from floor level. This appeared to be due to secondary movement, since there was no change in mortar, and no break in construction could be followed round. A small cut-in socket on the south face probably related to the internal arrangements of the Southern Building.

The eastern buttress was similar in size and shape to the western, though its inner (west) side was slightly concave rather than convex. The south face was inclined at a uniform 25° to the vertical, and was formed of 'tumbled-in' stonework at 90° to this; the junction between this and the horizontal coursing of the rest of the structure was awkwardly formed, but did not appear to be a rebuild. At its base, just above floor level, it overhung a vertical face, seemingly continuous with the masonry of the ashpit. The inclined face ended upwards at a slight ledge just below the outermost lintel of the arch, but the line of the buttress was continued upwards for 0.33 m from this, as a block of slightly-projecting vertical masonry separated by a straight joint from the main wall face.

The sidewalls of the arch rested on the walls of the ashpit (which is described below as part of the internal structure), leaving ledges 500–300 mm wide; these continued under the face of the arch walls to form rough horizontal slots up to 10o mm deep and 140 mm high, partly infilled with small stone, brick and firebrick fragments set in crumbly white sandy mortar. At the inner end of the arch, a single surviving iron plate (25 mm thick) across the top of the ashpit was bedded into these slots, indicating their function. Above these slots, the only features within the sidewalls were two pairs of adjacent cut-in sockets set opposite each other near the inner end, and a wooden wedge and an area of patching of no apparent significance.

The inner parts of the sidewalls directly underlay the vault of the arch, but the outer parts terminated at the base of horizontal recesses built into the buttresses, and ending at a vertical east-west face across the top of the arch, beneath the outermost lintel. These outermost lintels consisted of ferrous bars 35 mm square, their ends resting precariously on brickwork, here partially infilling horizontal sockets 150 mm high in the sides of the arch; this suggests that the lintels had replaced larger stone or timber lintels, for which the recesses had been designed. One lintel was sampled, and shown to consist of wrought iron (*see Ch 7*). The brickwork, and the masonry immediately above the lintels (on the line of the main south face of the furnace), was re-bedded in powdery fine sand with mortar lumps; this extended up to 0.5 m above the ferrous lintels, and presumably resulted from their insertion. Behind the lintels, it blocked the opening of a steeply-angled shaft leading up to the interior of the cone.

Beneath the lintels, at both ends, small rectangular openings ('bar-holes') entered the furnace structure, opening internally just above the tops of the cementation chests (*see below*). These were 230 mm wide and 280–310 mm high, and were built of firebrick set in pink-red sandy mortar, with one local rebuild in Type IV mortar. The structure of the holes appeared to be inserted, probably as a matter of routine replacement. The floors sloped down to the outside, and were heavily worn; this slope had been continued by the base of the recesses along the sidewalls of the arch, of which a single deeply-worn sandstone block survived.

The space between the bar-holes was occupied by the outer faces of the voussoirs of the main arch (the crown of which lay 0.4 m below the outermost ferrous lintels), the infill above these consisting of firebrick and stone pointed and patched with Type IV mortar. The arch form approximated to a shallow triangle, of two almost-straight limbs pitched at 10° to the horizontal; the voussoirs consisted of cut firebricks (some machine-made), set in brown sandy mortar. The arch sprung from the tops of three wrought iron lintels (77 mm wide by 16 mm thick), the outermost of which had corroded through. The space between the surviving lintels and the arch was packed with firebrick and sandstone set in the same brown sandy mortar as the arch. The lintels in turn rested on single courses of firebrick, set in brown sand and lime (probably a decomposed mortar), resting on shelves along the tops of the stone sidewalls of the arch.

The arch structure just described extended for 0.4 m inwards. Behind this, the innermost part was lintelled with large sandstone slabs, 0.2 m lower than the crown of the firebrick arch, each stone lintel being itself supported by a bar-like iron lintel (65 x 20 mm in section) beneath its axis, bedded into the original masonry of the sidewalls; the inner of these iron lintels was inset into the base of the stone lintel. The undersides of the sandstone lintels were severely reddened by heat.

All the features of the arch so far described formed parts of the outer casing of the furnace. However the rear face of the arch was structurally part of the interior of the furnace, forming the ends of the firegrate and chest compartment described below (*Ch 5*). Brick walling at the inner ends of the sidewalls also related structurally to the rear wall rather than the furnace casing.

The inner sandstone lintels of the arch were abutted by a segmental arch (80 mm high from springer to crown) of firebrick, laid alternately on edge and on end, and bedded in hard pale brown sandy mortar; its inner side abutted the ends of the chests and the end crosswall between them. The underside of this arch was infilled by a wall of header firebricks, 240 mm thick and 200 mm high at the ends, its top rubbed to fit the arch, abutting the chest structure to the north and exposed to the south. This rested on a wrought iron lintel 11 mm wide and 25 mm thick, bedded into the brick sidewalls.

Below this level, much of the rear of the archway was occupied by the open cavity of the firegrate, with narrow exposed strips of the ends of the chests and of the firegrate walls on each side, and the base of the crosswall above. These were covered by rough unburnt mortar, and it was clear that some structure in front of them had been lost. The inner 8 mm of the brick sidewalls were also 'raw', with a vague vertical boundary to the weathered faces to the south.

The inner ends of the sidewalls were composed of firebrick bedded in sandy lime mortar; the surfaces were burnt, but not vitrified. The brickwork was abutted by the ends of the chests and the firegrate to the north, and abutted cut-back ends to the sandstone sidewalls to the south. On both sides, it was largely broken away below the level of the 'chest flues' (*see below*).

The north face of the furnace was very similar in general form to the south face. The two end portions (outside the buttresses) were in alignment with each other, and the first and third plinths were consistent along the straight lengths. Several slight variations in build were examined carefully on site, but are not now considered significant; there was considerable sub-vertical cracking, most cracks being at least partially packed with stone and mortar. The upper part of the wall was cantilevered out from the straight to a bowed line, as on the south face. The central part, above the outermost lintel of the firegrate arch, was marked off by wide vertical cracks at each end; these had pulled open along two unusually straight joint lines, possibly a deliberate feature to allow for replacement or differential thermal movement of this area. The second plinth was absent, but at the east end there was a slight built-in shelf 0.3 m above its normal level, and at the west end there was a chiselled-in horizontal rebate at the same level. These features were at the same height as the mortice on the beam against the north-west corner of the furnace (*see above*), and as the tops of the two buttresses.

The eastern buttress was also very similar to those on the south face; it was formed of horizontally-bedded masonry throughout. The lower west face overhung considerably, and may have been rebuilt. The western buttress, however, differed in several details; its outer (west) side was at 95° (rather than 90°) to the main wall of the furnace, it was 1.8 m long (rather than 1.5–1.6 m), and the inclined north face was formed of 'tumbled-in' masonry very awkwardly joined to the remainder; it appeared to be a rebuild, and overlay the damaged top of the ashpit sidewall.

The outermost part of the arch was formed by a single ferrous lintel, 8 x 35 mm in section, holding the bowed centre of the main face; the west end of the lintel rested on the top of the western

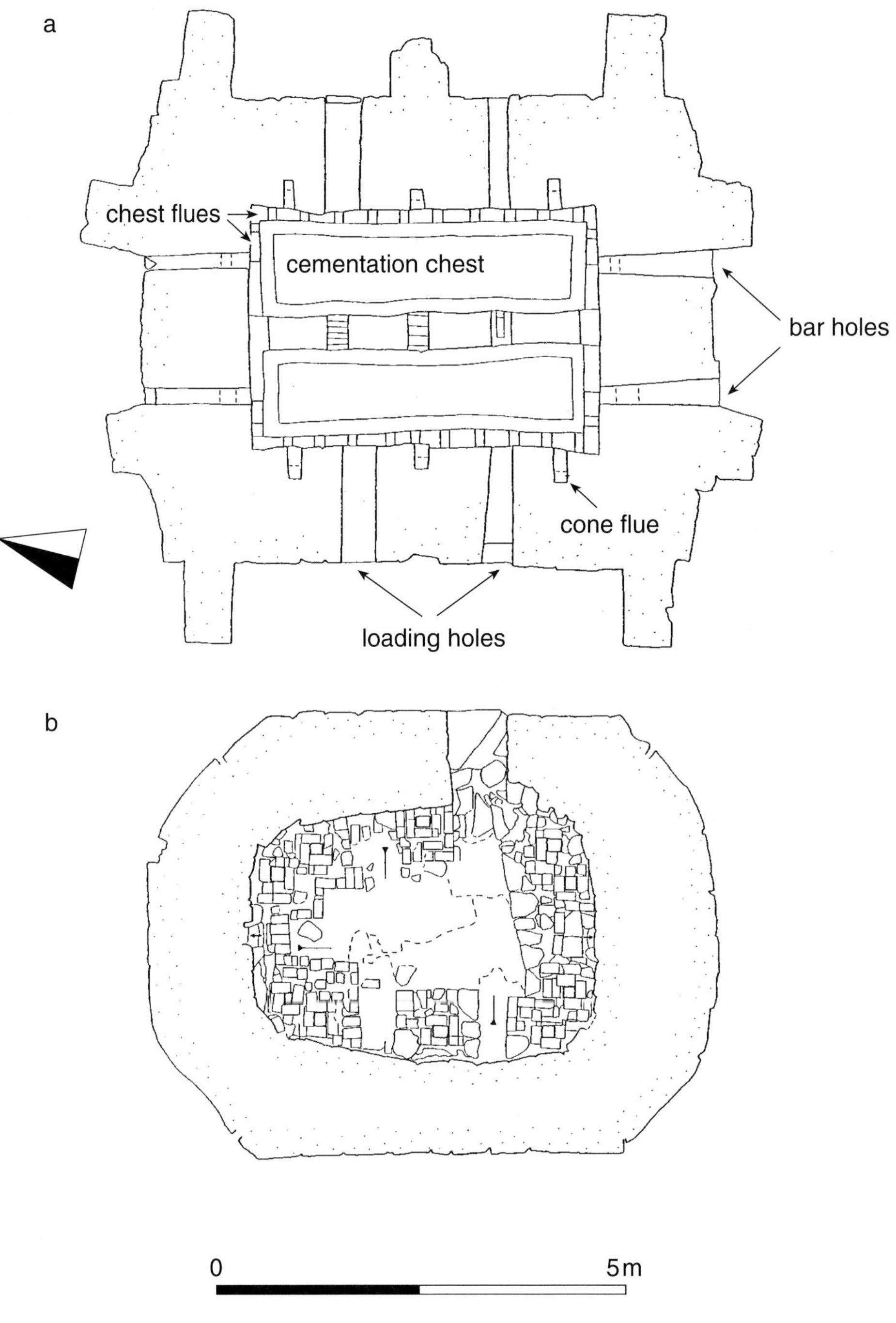

Fig 7   Plans of furnace: a) at chest level, showing the flue systems and loading holes; b) at cone floor level, showing the internal chimneys, and the mouths of the cone flues and the shafts from the firegrate arches

buttress, but the east end had rusted and moved completely off its former bearing, being supported only by a tree-stump anchored in the bar-hole below it. The lintel was sampled for metallographic analysis, and shown to be high-quality wrought iron (*see Ch 8*). Behind the lintel, a shaft 300 mm square penetrated from the archway to the cone interior, running upwards at an angle of 35° to the vertical. This opened from the centre of a 0.6 m height of sandstone walling, extending from immediately behind the ferrous outer lintel down to the stone lintel of the inner part of the archway.

This walling also formed the rear faces of two recesses along the inner sides of the buttresses. The eastern of these was very similar to those on the south face of the furnace, and the walling just below the lintel here contained a bar-hole 330 mm square (tapering inwards to 200 mm square), built into the original stonework, with a worn stone base. This was infilled with pink sand and mortar, and had probably been blocked off externally (the presence of a tree stump growing from it precluded certainty). The west side was however more complex. An original opening 200 mm square, matching that on the east side, had been blocked with firebricks in brown sandy mortar. This mortar continued down the wall face, infilling a crack, and was continuous with the bedding of a lower, unblocked opening (210 mm wide by 150 mm high) built entirely of machine-made firebrick (Plate 14). This was clearly inserted, and the recess in the side of the buttress had been cut down and patched to give access to it. Internally, this opening gave access to the base of the western cementation chest.

The remainder of the firegrate archway was simpler than its equivalent on the south side. It was roofed with sandstone lintels of varying width, bedded in Type III mortar, resting on a series of four flat iron lintels, varying in cross-sectional dimensions; these latter were very corroded, and some had broken off. The innermost was inset into the base of the overlying sandstone lintel. The lintels rested on the original sidewall of the arch to the east, and on the machine-made firebrick of the inserted lower bar-hole to the west (this latter relationship probably being produced by underpinning). The level of the lintels dropped slightly to the south, by 6 mm in the total length of 1.4 m.

The rear wall of the archway was also similar to (and less well preserved than) its southern equivalent. The top of the wall had been formed by the outer face of the end cross-wall between the chests, but this had been broken away below the level of the archway lintels. Its east end had abutted the end of the chest sidewall; the west end had abutted this internally,

but its outer side had abutted an infill of refractory sandstone filling a 0.18 m gap beyond the end of the chest, though still resting on its base slab. The outer faces of this infill, the cross-wall, and the firegrate walls formed a rather irregular vertical joint between the internal structure and outer casing of the furnace.

Immediately outside this joint, the archway was spanned by a low segmental arch of firebrick voussoirs, 230 mm wide, its outer side being separated by a further joint from the sandstone lintels of the arch described above. The east end of this rested on an area of firebrick sidewall, abutting the (ragged) broken end of the sandstone arch sidewall. The west end rested (probably due to underpinning) on the machine-made firebrick of the inserted bar-hole; below this, the inner end of the west sidewall was formed of similar machine-made firebrick walling, forming a vertical strip 300 mm wide against a vertical stopped end to the original sandstone masonry to its north. The surface of the brickwork, on both sides, was surprisingly fresh-looking, with no sign of the severe heat damage expected in this location. It had probably been masked in use by some structure now robbed from its front, but no clear scars to confirm this could be seen.

The cone of the furnace (Plate 15) was oval in plan; it measured 7.0 x 5.3 m externally at its base, tapering to an almost-circular 2.5 m at its broken top (*c* 8 m above ground level). It was constructed of coursed dressed sandstone bedded in Type III mortar, the walls tapering upwards from up to 1.4 m thick at the base to 0.6 m (measured on the horizontal) for most of its height. No evidence for the original top survived; the remainder of the structure was intact apart from numerous sub-vertical cracks, some of which were packed with mortar and stone or firebrick pieces.

On the east and west sides, the junction between the cone and the lower rectilinear structure was formed by a curving arris, dipping to third plinth level at the corners. On the north side, it was formed by 0.2–0.4 m of curved and battered wall above the third plinth, separated by a ledge of variable width from the cone proper. On the south side, however, the central part of the cone face was carried down in a convex curve to a vertical face above the third plinth, while towards each end flat-topped triangular platforms were introduced to fill the gap between the two plan-forms.

The only opening in the cone structure lay on the south-east quadrant, 0.5 m above one of these platforms, and just above the verge line of the Southern Building roof (*see below*). This opening was 0.34 m wide and 0.68 m high; both floor and roof were stepped up inwards. The opening was blocked internally by a firebrick patching to the inner face; outside this it was partly infilled with earth and rubble, but was open to the exterior. Directly opposite this, in the north-east quadrant, a major crack had followed an apparent stopped end formed by superimposed joints in two courses of masonry. This may possibly have been a vestige of a similar opening; however if so it must have been abandoned during the initial construction of the cone.

The only remaining features lay on the northern and southern quadrants, and formed traces of the verge lines of the Northern and Southern Buildings respectively. On the north side, this was marked by a band of intermittent adhering mortar, with patches of slight damage or cutting-back of the stonework; no identifiable beam sockets were present. On the south side, the lower west pitch of the roof was marked by a band of adhering hard white mortar, replaced upwards by an incised groove up to 120 mm wide and 50 mm deep, at an angle of *c* 40° to the horizontal. The apex was marked by a further patch of adhering mortar, but on the eastern pitch only isolated fragments of mortar survived. Below the verge line, a scattering of cut-in beam sockets was present, varying in reliability of identification; they were between 100 mm and 400 mm below the verge line. A socket at the apex had presumably held the ridge-piece of the roof.

The interior

As already stated, the interior of the furnace was divided into three major spaces: the ashpit and firegrate, the chest compartment, and the cone interior (Fig 7). The ashpit was exposed by excavation (being totally infilled by post-abandonment stratigraphy), and some excavation was also performed in the cone interior. The former excavation is described in Chapter 6, but the latter is included in the present section, since the excavated deposits related closely to the solid structures. The drawn record was prepared by hand, and much of the recording had to be done under artificial light, due to the enclosed nature of the spaces.

The ashpit lay along the central north-south axis of the furnace. Its central part was parallel-sided and 0.5 m wide, but the two ends (beneath the firegrate arches) widened to 0.9 m at the base, and 1.1 m at the top (the sidewalls being slightly battered back from the vertical). Access was by flights of steps forming the north and south ends of the structure. The ashpit was 8.0 m long at the base, and 10.0 m long to the top of the steps; it was 1.2 m deep from the floors of the Northern and Southern Buildings, and 1.45 m from the line of the actual grate.

The floor of the ashpit was composed of rectangular sandstone slabs 300–500 mm wide, laid across the pit and passing under the bases of the sidewalls. It sloped down very slightly from both ends towards the middle. There was an area of heavy wear under the south end of the firegrate, where one slab had apparently been replaced. At the south end, the original floor was overlain by a replacement of thin stone slabs set in hard sand and crushed sandstone, abutting the walls and the bottom step. This may have been inserted to raise the level, since the original floor beneath it showed no signs of severe wear.

The steps at the south end (Plates 16, 17) were formed of long sandstone blocks, with smaller coursed masonry beneath them on the risers. The risers averaged 240 mm high, and the treads 220–280 mm wide. There had originally been five steps, but the top two had been combined, by the insertion of a secondary step of housebricks set in brown mortar and earth; the combined step survived to a height of 370 mm, the top being broken off. This alteration may have been to accommodate a circular cast iron plate in the floor of the Southern Building (*see Ch 6*). The northern steps had been similar in their original form; here the third and fourth steps (from the bottom) had been combined by the insertion of a massive sandstone riser over the original third step; a mortar scar indicated the site of a robbed top step over this. Beneath the bottom step, a culvert 360 mm wide and 170 mm high opened into the pit, the floors of the ashpit and culvert being continuous. This culvert continued beneath the Northern Building and its northern buttress, and was located in the excavation beyond this (*see Ch 6*). It presumably functioned as a drain, and perhaps also as an air inlet for the draught of the furnace.

The sidewalls of the ashpit were formed of roughly-coursed sandstone masonry bedded in hard white Type III mortar. Their condition was good near the ends, but the stone was severely reddened and eroded under and close to the firegrate, and the walls had been broken off upwards and replaced by the brickwork of the firegrate structure (at an irregular height of between 0.22 m and 0.6 m above floor level). A few small sockets were cut into the faces of the sidewalls, but these could not be interpreted.

The tops of the ashpit walls were separated from the bases of the buttresses by distinct shelves, varying in width from 2 mm to 300 mm (widening overall outwards), and extending into ragged horizontal recesses beneath the faces of the buttresses. Most of these recesses were infilled with earth and mortar debris, but at the inner end of the southern firegrate archway an iron plate survived, set into the recesses and extending across the width of the ashpit. This rested on a strip of brick and stone pieces in the base of the recesses, which abutted the face of the firegrate; its inner edge was also supported by a very-corroded wrought iron lintel, consisting of a bar c 5 mm in diameter, over a plate 160 mm wide and 3 mm thick.

At the north end, the top of the western sidewall was broken away, and overlain by a rebuild of the western buttress (*see Ch 5*).

The firegrate was continuous downwards with the central part of the ashpit, and opened upwards into the centre of the chest compartment, its upper sidewalls being formed by the walls of the chests. It was divided by a slight horizontal shelf along both walls; below this the wall faces were heated but undamaged, whereas above it the faces were intensely vitrified by heat. This shelf can therefore be reliably interpreted as marking the level of the grate itself.

The lower walls of the structure were built of coursed firebrick; the faces were slightly inclined inwards, the cavity declining upwards in width from 0.48 m to 0.30–0.34 m (at grate level, where the walls were vertical). At least three builds were present. The earliest (surviving in the lower and northern parts) consisted of sand-faced firebricks measuring 250 x 120 x 60 mm, bedded in red to grey sand. The second phase consisted of smooth firebricks (240 x 120 x 70 mm) with sharp arrisses, bedded in pale red sandy mortar; it formed a rebuild, largely on the east side. The final phase, forming the upper southern part of the west wall, consisted largely of smooth sharp-cornered white firebrick, but included tile-like pieces and one large slab; it was bedded in hard cream-white sandy mortar with pebble inclusions. The faces of all phases were reddened by heat, with some slight vitrification at the top.

The firegrate space was divided across its centre by a crosswall, resting on a composite lintel of iron bars and rods, 1.35 m above the floor of the ashpit. The wall itself was 340 mm thick and 150–400 mm high, its top being concealed within the sheet of vitrified ash above the grate (*see below*); it was composed of coursed firebrick. The lintel was slightly wider than the wall, forming a slight shelf on each side.

The space was also spanned by a number of iron bars. Seven of these (three to the south of the crosswall, and four to the north, two forming an adjacent pair) were set at the same level as the lintel of the crosswall, at intervals of 500–600 mm apart; the majority were square bars set diamond-wise, but their detailed form varied. They were 6 mm below the ledge at the (abrupt) base of the intense vitrification, and are interpreted as the supports for removeable longitudinal firebars. The remaining two bars were set at a slightly lower level, near the outer ends of the firegrate; they may have been structural braces, or pivots for a poker.

The upper part of the firegrate structure was formed by vertical sidewalls, offset slightly back from the lower walls by the ledge mentioned above. However these upper walls differed markedly in appearance, being covered by a single sheet of purple vitrified material (which had obscured most of the structural detail). This sheet extended across the northern interior of the area at the level of the ledge, as a self-supporting sheet up to 300 mm thick, adhering to the central crosswall but suspended 6 mm above the iron bars. A similar sheet had probably existed in the southern half, but had broken away.

Above the ledges, the wall was formed by six courses (430 mm) of firebrick, bedded in a hard crumbly cream-white material (probably a fireclay luting). On the east side, the wall was laid in English bond, while the west side was less regular. Where the walls were damaged, they could be seen to have rear faces, abutted by an infill of mortar, brick, and sandstone rubble beneath the cementation chests. The outer faces were completely coated with vitrification; this averaged 3 mm thick at the base (where it adhered to the undamaged faces of the bricks), thinning to 1 mm at the top (where it had eaten into the surface of the bricks). The sheet tended to form humps and hollows below the mouths of the 'chest flues' (*see below*).

Immediately above the walling just described, a set of horizontal flues opened from the firegrate space; above these the sides of the space were formed by the inner walls of the cementation chests. This area was therefore structurally part of the chest compartment, but was functionally part of the firegrate area.

The flues varied in detailed form and dimensions, averaging 200 mm square in section, and 200 mm apart; there were eleven on each side, but they tended to be staggered rather than opposite each other. The flues ran horizontally back from the firegrate for 1.2 m, then turned vertically upwards. They were divided from each other by single blocks of fine-grained cream-white sandstone ashlar laid east-west, each block measuring 1200 x 200 x 200 mm. Their openings to the firegrate were partially infilled with vitrified debris, merging back into un-vitrified dust. The north end flues on each side abutted the internal face of the brick arch structure at the rear of the firegrate arch, and also fed into a central flue running up the end of each chest; this latter feature was repeated at the south end, where the flues did not adjoin the arch structure.

The roofs of the flues were formed by massive horizontal sandstone slabs, 200 mm thick, extending the full 1.2 m length of the flues. So far as could be seen, each single slab also extended the full 4.3 m length of the chests.

Above these slabs, the top 0.92 m (3 ft) of the sidewalls was formed by the walls of the cementation chests. The surfaces were vitrified into a continuous mauve crust, less glossy than that of the lower walls. Where this was damaged, it was seen to form a luting (presumably of fireclay) 10 mm thick, over the face of the actual walls. Where visible, the east wall was composed of white crystalline sandstone ashlar, in courses 150–200 mm high. However, where the composition of the west wall was visible, it consisted of firebrick slabs in uniform 230 mm courses.

Between the chests, the central space was spanned by five crosswalls, each c 240 mm thick; these were supported by low arches springing from just above the 'chest flues', and had tops at or just below the rims of the chests. Their form differed in detail, but all were constructed of firebrick, and their surfaces were severely vitrified; some had cracked or distorted during use, before the final vitrification. The two end crosswalls abutted the inner faces of the chest compartment vault, above the inner ends of the firegrate arches. The northernmost wall had suffered considerable recent damage, revealing that it was constructed of cream-coloured firebricks stamped 'RAMSAY', bedded in cream-coloured fireclay.

The firegrate opened upwards into the centre of the chest compartment (Fig 8(a), Plates 18, 19) This formed a rectangular vaulted chamber, measuring 4.3 x 2.8 m at the level of the chest tops, and 1.2–1.3 m high from this level to the crown of the vault (which had sagged slightly in the centre).

The lower part of the compartment was totally occupied by the two cementation chests, and the top of the firegrate space between them. The chests were not identical in form. The western chest measured 3.60 x 0.78 m internally, by 0.90–0.94 m deep. The east and west walls were 160 mm thick, and the north and south walls 180 mm; the east wall was considerably distorted and cracked. The inner faces and rims were luted over with fireclay, so that the structure was rarely visible; where it was, the west wall was composed of grey sandstone, and the east wall of firebrick slabs 450 mm long. The internal luted surfaces were dark grey below a line of adhering vitrified material 0.70–0.74 m above the base, and dull red-brown above this. The base consisted of a single sandstone slab.

The north end of the chest contained a single opening, 130 mm wide by 70 mm high, the sill being 330 mm above the base of the chest. This formed the inner end of the inserted lower bar-hole, the outer end of which has been described above (see Ch 5) in the northern firegrate arch. Over and around the opening, the chest wall had been patched with firebrick, covered with a fresh-looking luting. This and the original luting were both reddened from the base of the opening upwards. Both lutings were marked by a scatter of horizontal dark iron-stained dents (the clearest measuring 7 x 2 mm), probably produced by bars touching the end during a firing.

The eastern chest measured 3.60 x 0.69 m, and was 0.92–0.96 m deep. The walls were 160–180 mm thick, and the west wall leaned slightly inwards. The luting increased in thickness downwards (to 5 mm at 200 mm below the rim; where visible, the walls behind it were composed of grey sandstone. Remaining features were very similar to the western chest, except that there was no opening through the walls of the chest.

The outer sides of the chests were separated from the walls of the compartment by the upward openings of the 'chest flues'. These formed a continuous series of vertical openings, 200–340 mm long and 140–200 mm wide, separated from each other by walls of firebrick one brick (110 mm) thick. They extended downwards for 1.32 m, opening into the tops of the (narrower) horizontal flues beneath the chests (see above). At the ends of the chests, there was one flue opening in each corner of the compartment, and one beside this between the ends of the chests and the end walls of the compartment; the inner parts of the chest walls were separated from the compartment walls by solid infill of firebrick or masonry.

At the level of the chest tops, the walls of the chamber were straight and vertical (except for local distortions). The curvature of the vault began 200 mm above this level; the east and west walls were carried over as a single barrel-vault, while the north and south walls were curved in to form segmental ends to the vault, the tops of which were curved in. The vault was composed of firebrick headers (15 x 7 mm in section), their faces vitrified to a purple-black glassy consistency. In places the brickwork was coated with a red-brown 'biscuit-like' luting of fireclay, probably the result of running repairs between firings. The whole surface was irregular in detail, with many vitrified cracks, and had also distorted and sagged overall. In two areas, the surface had flaked off before or during the final use. The vault also contained several areas of more substantial patching, identified by their fresher and less vitrified appearance. These

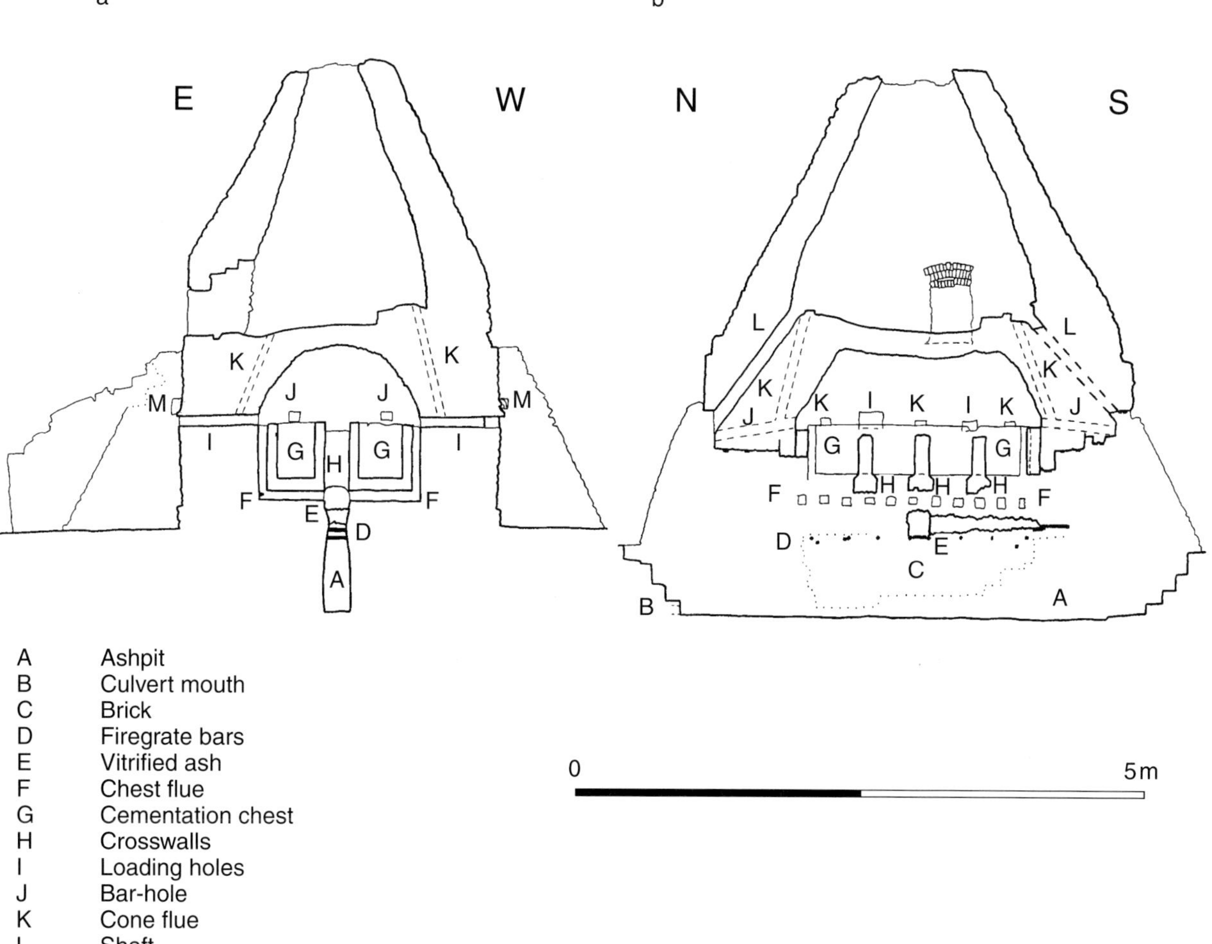

A	Ashpit
B	Culvert mouth
C	Brick
D	Firegrate bars
E	Vitrified ash
F	Chest flue
G	Cementation chest
H	Crosswalls
I	Loading holes
J	Bar-hole
K	Cone flue
L	Shaft

Fig 8   Cross-sections of furnace, to show the internal structure

tended to occur round the mouths of the 'cone flues', but varied considerably in extent.

The walls of the chest compartment contained a number of small openings, all close to the springing of the vault, with their floors at or close to the rims of the chests.

The east and west walls were each penetrated by two 'loading holes' penetrating to the exterior (where their openings have been described above (*see Ch 5*)), the northern hole on each side being larger than the southern. In the east wall, the southern loading hole was 240 mm wide, and had been 190 mm high before the lintel had sagged. This lintel consisted of a slab of firebrick or sandstone measuring 450 x 220 x 100 mm, behind which the roof was of rough sandstone. The walls were of vitrified coursed firebrick stretchers for 360–480 mm from the inner opening, and of sandstone rubble outside this. The floor was obscured by vitrified material. The northern loading hole measured 420 x 300 mm internally. It was lintelled throughout with firebrick slabs (500 x 460 x 80 mm),

and was walled with firebrick. The floor consisted of vitrified luting or dust, sloping outwards, with rough sandstone rubble outside this.

In the west wall, the southern loading hole was 250 mm wide and 210 mm high. The walls and roof were of firebrick internally, and sandstone outside this, while the floor was of vitrified luting, probably over firebrick. The opening was blocked externally, and was severely vitrified internally; unlike the other openings (all of which had been replaced or repaired), it was primary to the existing vault. The northern loading hole measured 420 x 300 mm, and was lintelled throughout with firebrick slabs (520 x 460 x 75 mm). The walls were formed of firebrick, and the floor of vitrified luting or dust internally, and rough sandstone outside this.

The south wall contained the internal openings of the two 'bar-holes'. The western of these was 200 mm wide and 150 mm high, and was walled with firebrick. It was lintelled internally with a firebrick slab, behind which a cone flue (*see below*) opened into the roof.

There was a butt joint in the walls at this point, outside which the whole structure sloped gently down towards the exterior. The eastern hole was 200 mm square, and was walled and lintelled with ordinary firebricks (one stamped 'RAMSAY'). A cone flue opened from the roof, and the whole structure sloped slightly outwards.

The north wall also contained the inner openings of two bar-holes, though these had been blocked externally (*Ch 5*). The eastern of these measured 160 x 90 mm, and was blocked by a solid mass of luting or mortar 0.64 m from its mouth; a cone flue (not blocked) opened from its roof. The western hole measured 160 x 110 mm, its base having been raised by 60 mm from its original level. The hole dipped outwards, and was blocked by rubble 1.20 m from its mouth. A cone flue opened from its roof.

The remaining openings in the springing of the vault were smaller than those so far described, and did not penetrate to the exterior; instead, they turned vertically upwards, and vented by small internal chimneys (*see below*) into the interior of the cone. There were three of these 'cone flues' in each of the long walls of the chest compartment; as already described, the two cone flues in each end wall opened from the tops of the bar-holes. The flues were lintelled at their mouths with single firebrick slabs; many were encrusted with a hard black sooty deposit, the surface of which was marked by ripples converging into the flue mouths, indicating the flow of hot gasses. The flues differed in precise dimensions; all were around 180 mm wide, but the height at the mouth varied from 70 mm to 220 mm, the sills of most having been raised during use to reduce their height. The flues extended horizontally for 0.25–0.44 m, before turning vertically or very steeply upwards.

Of these openings, the bar-holes and flues were clearly far too small for man access; so were the southern loading holes in each side. The enlarged northern loading holes would just allow a very small adult to crawl in, but (as today) the easiest access was probably by climbing from the firegrate into the centre of the compartment.

The interior of the cone formed a relatively simple enclosed space (Fig 8(b), Plates 20–22). At floor level it measured 4.2 x 3.0 m internally and its plan form was rectangular with rounded corners and bowed sides. It tapered upwards into an ellipse, measuring *c* 1.2 x 0.8 m where it was broken off (at 4.0–4.5 m above floor level).

The wall of the cone was formed of coursed sub-rectangular sandstone masonry set in pink-white mortar. The extreme base, below a slight projecting shelf, was almost vertical and the face was undamaged. Above the shelf the wall was corbelled in at varying angles to form the cone, the junctions between the angular lower corners and the rounded cone itself being formed by cantilevered overhangs. Much of the surface was covered by a hard friable crystalline crust of black-stained material, and the stone was reddened and friable from heat in places.

At the base, the centre of each end of the cone contained the top of a rectangular shaft, measuring 300 x 250 mm and communicating downwards at a steep angle to the northern and southern firegrate arches (the shaft at the south end was blocked, with a rubble infill almost to its top). The lower part of the shafts were entirely built into the structure of the outer casing of the furnace, whereas the top 600 mm was formed by a recess in the inner face of the casing, separated by a narrow airgap from the steeply-battered outer sandstone wall of the chest compartment vault. The shafts terminated upwards at the shelf in the wall, which projected beside them.

The only open access into the interior was by the opening in the base of the east side, south of centre (the exterior of which is described in Chapter 5). Behind the external lintel, the head of the opening was formed by a segmental arch of edge-laid firebrick, with rubbed wedge-shaped firebrick keystones. The firebrick was set in three crossways rows, each set 120 mm higher than the one to its outside. The sides of the opening were formed of coursed sandstone masonry. Behind the external sill, the floor consisted of solid flat sandstone rubble set in hard pale-red sand; this continued internally beyond the inner face of the wall, but was only exposed in a test trench.

To the south of this, the blocking of the small enigmatic opening (*Ch 5*) was visible; this consisted of coursed firebrick bedded in soft white clayey sand. The inner face was vitrified by heat, indicating that it had been inserted before the final use of the furnace.

The features occupying the floor of the cone interior were exposed by excavation; since most of the layers investigated were interstratified with the structures, both layers and structures are described here. The outer face of the chest compartment vault was not exposed, even in a test trench in the doorway in the east side. This test trench was terminated at the surface of the solid flat sandstone rubble described above, the inner part of which must have overlain the vault (directly or indirectly). The rubble was overlain by a very hard purple-red sand containing mortar or firebrick fragments and burnt sandstone pieces (only partly removed). On the north side of the test trench, a footing of better-laid rubble, within the layer, supported an internal chimney (*see below*), while to the south an area of sandstone walling simply rested on the surface of the layer. Between these, the hard sand was overlain by a further very hard light pink-red sand, up to 100 mm thick, abutting the faces of the walling and the internal chimney. This was only removed in the test trench.

The south end of the interior was occupied by an area of sandstone masonry, reddened by heat, set in pink-white mortar, with a roughly-coursed face to the north; this ran slightly south of west from the south side of the doorway. The masonry stood 200–350 mm above the base of excavation to its north, the latter height perhaps being original. To the south, it underlay the internal chimneys and some brickwork between them; it may have been continuous beneath the latter with some angled slabs forming the north side of the blocked shaft in the south end.

At the north end, the inner face of the shaft was formed by a face of steeply-battered coursed sandstone, which continued to both east and west, separated from the cone wall by a narrow void; the bases of the internal chimneys were built into it. The top of this walling was obscured by the hard sand surface of the central interior; however it projected above the level of much of this surface, and cannot therefore have been the external face of the main vault structure. Intermittent stones in the surface beside the east and west walls of the area may have been its continuation.

The dominant feature of the interior, however, was a set of internal chimneys, projecting from the hard earth base. There were six chimneys, containing the orifices of the ten 'cone flues' whose lower parts have already been described; chimneys in the four corners of the iterior each contained two flues, whereas chimneys in the centre of the east and west sides contained a single flue.

The chimneys were faced with firebrick and common brick set in hard red-white mortar; some had a separate core of firebrick and sandstone rubble. They varied in detailed shape and size, most approximating to a 1.1 m square or 'L' shape. The tops of most were broken off, typically at 150–350 mm above the hard earth surface. The south-west chimney reached 0.4 m above this surface, and its top was vitrified. However this top was composed of mortar, suggesting that one further course had been robbed off. The flues were typically c 160 mm square, and were steeply angled rather than vertical. Their inner faces were vitrified into continuous sheets, broken off at the tops. The eastern flue in the south-east chimney was less severely vitrified, and differed in detailed form; it was probably a late repair. It was composed of stamped 'RAMSAY' firebricks.

The internal chimneys were normally separated from the cone wall by slight gaps, except along the east side (where they abutted the wall), and in the four corners; here they extended into the recesses formed where the curving cone wall was cantilevered over the corner of the underlying rectilinear walls, so that they appeared to underlie the cone walls locally.

Between the chimneys, the base of the area was formed by a consistent hard rammed surface, varying in composition from bright red sand with some firebrick and stone, through pink sand with burnt sandstone, to an off-white mixture of sand, firebrick rubble, and decomposed firebrick. The surface dipped from the sides to the centre, and from the centre to the door.

This surface was overlain by an average 0.4 m of less-solid earth stratigraphy, which was removed. The lower layers consisted of a low dome of red-grey sand in the centre (perhaps from the weathering of the friable inner face of the cone structure), with varied deposits of sandstone and firebrick rubble and sand lenses round the sides. These were overlain by softer and darker deposits, encroaching across the broken tops of the chimneys; the surface of this sequence contained recent material. It was overlain by up to 0.1 m of loose dust, earth, and rubble.

The Northern Building

The Northern Building was a slightly irregular rectangle, measuring 6.3 m (east-west) by 4.3 m (north-south) externally. The roof had been gabled, the ridge running from the north wall to the cone of the furnace. The north (gable) wall survived virtually intact, and the west wall was largely intact except for its upper north end. The east wall, however, had collapsed to c 1 m above floor level. The single doorway had always been at the north end of the east wall, and the building showed few signs of alteration, except for the addition of a large buttress against the external centre of the north wall.

Like the Southern and Eastern Buildings, the Northern Building was constructed of sandstone rubble (softer and browner than that used in the furnace), bedded in Type I 'mortar' (actually a brown sand and silt), with traces of Type V mortar pointing surviving on the less exposed external faces. The core, where visible, was of small sandstone pieces bedded in brown sand. Projecting throughstones were rare, and are noted below; the frequency of non-projecting throughstones could not be assessed. The wall faces were roughly coursed, the rubble being predominantly of sub-rectangular blocks; however many walls also contained irregular pieces of stone, some laid with their bedding planes close to the vertical.

The east wall stood to between 0.8 m and 1.2 m above floor level, its upper courses leaning and bulging increasingly outwards. It was 0.6 m thick, and much of its Type I bedding had been converted to soil by root action. The external junction with the furnace was pointed with Type V mortar (which may well have been weathered off elsewhere), and part of the internal face retained a rendering or repointing of hard pale grey lime or lime-cement mortar. The inner face abutted the north wall of the furnace, but the outer face passed outside the corner of the furnace, abutting the northern buttress of the furnace east face (0.2 m back from the corner of the furnace). It had been keyed to the furnace by two projecting tusks and one recess forming a 'negative tusk', all on the north face of the furnace.

The wall terminated to the north in a stopped end, forming the south side of a doorway 1.15 m wide. This doorway retained a stone sill, built into the bases of the east and north walls, and containing the 2 sockets (each 140 x 120 x 20 mm deep) for a doorframe. An iron spike had been driven into the stopped end of the wall, 0.4 m above the outer side of the adjacent socket; this spike was 22 mm square and projected by 78 mm, with traces of a (broken off) head on its inner side. By analogy with the surviving door frame of the Southern Building (*see below*), it probably held a scarfed-in replacement for the base of the frame.

There was no surviving evidence for or against a window in the east wall.

The north wall was 0.6 m thick, increasing to 0.7 m at the east end. Several projecting throughstones were visible internally in the gable, and in the west end below eaves level. Breaks in construction were visible at eaves level (the gable being of rather larger blocks than the lower part), and 0.5 m below this (marked by a thin string course internally, and an unusually-consistent coursing line externally). The external corners were roughly quoined with larger blocks, improving in quality upwards. The wall stood to 3.3 m above floor level at the western eaves, and 5.65 m at the apex, both heights being probably only just short of the original (the top of its east end was more severely broken away). The broken top between apex and eaves was pitched at 38° to the horizontal; the outer face consistently stood around 0.3 m higher than the inner (respectively just above and just below the roof line as indicated on the face of the furnace cone), and retained a few neatly-tooled masonry blocks. This evidence, together with the

presence of abundant distinctive triangular ashlar blocks in the rubble below the wall, suggests that the gable had been finished off with a half-parapet of 'reverse crowstepping', a common local vernacular feature (Emery *et al* 1990, 132, and author's observation).

The internal face of the wall contained a row of three beam sockets at eaves level, 1.0 to 1.4 m apart. The sides of these sockets were damaged, but they had each measured around 200 mm wide, 250 mm high, and 200 mm deep. The south end of the beams may have rested on the plinth at the top of the north face of the furnace. Also at eaves level, the west end of the wall contained a horizontal beam slot, extending 0.65 m from the junction with the west wall, and holding the rotted traces of a beam; the height of this feature had been obscured by subsidence of the overlying masonry. At the east end, a noticeably flat top to the surviving wall face matched this precisely in level and location, and may have been the remnant of a similar slot. A further single slot was present on the axis of the building 0.5 m below eaves level, with no apparent lodging place for the beam in the face of the furnace opposite. The only other feature in the upper part of the wall was a single wrought iron strip or plate, driven at an angle downwards into the wall, 0.9 m above eaves level; its purpose was not apparent.

Two holes had been cut or broken into the internal base of the wall, just above floor level. The western of these consisted of a double hole, above a patch of brick flooring in the north-west corner, the lower hole opening to the exterior 0.3 m above external ground level. The eastern hole (180 mm wide by 200 mm high) had been blocked internally with brick and firebrick set in hard pale brown sandy mortar or cement, overlying the internal floor; this opened to the exterior just above local ground level.

The outer face had been pointed with Type V mortar, here containing fragments of pantile. It contained no features of interest.

The northern buttress abutted the centre of the north wall. It was 3.10 m long, 1.23 m wide, and 3.65 m high from modern ground level beside the north wall. It was composed of horizontally-bedded rubble (including one burnt, and therefore presumably re-used, block), with larger roughly-dressed blocks on the inclined north face (which was angled at an average of 38° to the vertical, curving to vertical at the base). The extreme top was broken off. Excavation revealed a slight a consistent slight plinth on both east and west faces, close to post-construction ground level beside the wall; a wider plinth at a lower level was exposed at the north end of both faces (due to the northward dip of the excavated surface). At the north end of the west side, the wall face between these plinths was chiselled back, possibly to form a rebate for a door into adjoining timber-framed building 1020 (*Ch 6*).

The buttress had been pointed with Type V mortar, and holes in its faces had been packed with pantile fragments (some bedded in the Type V mortar). The upper part of the butt joint against the north wall had opened to a gap 100 mm wide, the lower part being packed with loose Type V mortar and pantile pieces.

The west wall (Figs 4, 5, Plate 23) was of normal construction and thickness (0.6 m), with traces of Type V pointing. To the south of the window, there were remnants of external rendering, of a hard cement-based mortar similar to Type II, but containing sand and fine gravel. Its north end abutted the furnace, overlapping from the furnace north face to the northern buttress on the west face (in a similar manner to the east wall); there were two tusking stones on the furnace face, and one on the north face of the buttress. To the south, the wall stood to eaves level at 3.2 m above floor level, but over and north of the window it leant outwards, and parts of the top had collapsed.

The only opening through the wall was a centrally-placed window, 0.65 m wide by 0.86 m high. The sill was formed by the top of a course, with no special features, and the sides by simple stopped ends in the wall, with one quoinstone internally. The top was lintelled with beams, *c* 120 mm square, in the inner and outer faces, bedded in Type I brown sand. Between these lintels, the core of the wall was held by smaller timber lintels. All the timbers were severely rotted. The window frame survived, consisting of a simple rectangular frame of small timbers, each *c* 90 x 80 mm in cross-section, set back 140 mm from the external face and 350 mm from the internal face. The top and bottom members each contained two cylindrical sockets, 30 mm in diameter, spaced at 170 mm from each other and from the sides of the frame. These presumably held vertical wooden or iron bars.

The only features of note in the internal face were all 2.2 – 2.4 m above floor level, their bases lying at the level of the second plinth of the furnace north face. Two beam sockets were present in the intact northern part of the wall, one being at the north end and the other 1.25 m to its south. Both were filled by the rotted stubs of beams, 250 x 150 mm in cross-section, and were original features (their sides being formed by stopped ends, and their tops and bases by flat slabs built into the wall). The northern beam was laid flat, whereas the southern was laid on edge. Further

south, 1.25 m from the second beam, a flat stone in the broken wall top may have the remains of a similar feature. At the south end, the beam built into the buttresses of the furnace west face (*Ch 5*) continued for 0.34 m along the inner face of the wall, its end being embedded in the wall core. However since the wall had leaned outwards here, the beam may originally have been completely concealed within the core of the wall.

Features in the external face of the wall probably related to the adjacent timber structure discovered in excavation (*Ch 6*). At the north end, 1.8 m above modern ground level, a socket measuring 130 mm wide by 80 mm high and 110 mm deep was cut into two quoinstones of the corner; it was infilled with pieces of tile and firebrick set in mortar. South of the window, a patch of burnt stone set in the cement-based render probably infilled a similar socket, and near this a patch of firebrick set in the same cement may have infilled a horizontal slot along the wall face. Examination of this area was inhibited by the unstable state of the overlying masonry.

Two features in the west face of the adjacent northern buttress of the furnace may also have related to the same timber structure, since they lay directly above the excavated beam slot that formed its south end. The upper feature lay at a similar height to the socket at the north-west corner just described, and was a slot measuring 150 mm wide by 90 mm high and 150 mm deep. The second feature lay 0.9 m below this, and was a tapered hole measuring 230 mm wide, 100 mm high, and 150 mm deep; it may have been a beam socket, or merely a stone-loss hole.

The Southern Building

The Southern Building was a rectangular structure measuring 12.7 x 7.8 m externally, the long north-south axis lying at a slight angle to that of the furnace. It abutted the furnace to the north, and was abutted by the Eastern Building on the northern part of its east side. External access was originally by doorways north of centre in the west and east walls (the former having been blocked at a later date), and a second doorway in the east wall gave access to the Eastern Building. Before excavation, the east and west walls stood virtually intact to eaves level; the south (gable) wall was also almost intact, but was leaning so severely that controlled dismantling of its upper courses was necessary for safety reasons (after recording in as much detail as safely possible). The walls were similar in construction to those of the Northern Building (roughly coursed sandstone masonry bedded in Type I sand, with traces of Type V mortar pointing), but showed rather more evidence of alteration (including the total rebuilding of the south half of the east wall).

The west wall (Figs 4, 5) averaged 0.65 m thick, and stood 2.8 m high from internal floor level to eaves height (the original wall head surviving intermittently); the masonry included occasional water-rounded boulders internally, and the footing consisted of two courses of slightly-projecting rubblestone, the top of which lay close to internal floor level. It contained a (partly-blocked) doorway north of centre, and an altered window near the south end.

The abutment of the north end against the furnace differed from those of the east wall and the Northern Building, in that the west wall was carried entirely outside the corner of the furnace, butting against the southern buttress on the west side of the furnace. There was no visible tusking on this junction (and none on the corner of the furnace itself). The gap (60–150 mm wide) between the inner face of the wall and the corner of the furnace was infilled with coursed sub-rectangular sandstone masonry in Type I bedding, roughly keyed into the face of the wall; the infill may have abutted tusking stones projecting from the west wall, but the evidence for this interpretation was not convincing.

The doorway lay 3.0 m from the internal corner (3.6 m from the external north end of the wall); it was 1.9 m high and 0.98 m wide. The sides consisted of simple stopped ends to the wall, and the face of an original sill was visible externally, consisting of a pair of flat slabs built into the base of the wall. The doorway was lintelled by a row of timber beams, averaging 100 mm square, in Type I bedding; these had rotted and sagged.

The doorway had later been converted into a window, by the blocking of its lower part. The blocking consisted of a wall of coursed sandstone rubble and firebrick, with some brick and tile internally, and some used cementation chest sandstone externally; it was bedded in hard brownish-white sandy cement with small pebble inclusions, and rendered internally with a greyer cement covered by lime-wash. It was 0.84 m high, the external face resting on the original stone sill, and the internal face on a large stone lintel in the base of the blocking, over an iron tank set into the floor (*Ch 6*).

The space above the blocking was 0.98 m square, and was occupied by a timber window frame, set in firm sandy mortar with coal, animal bone, and pebble inclusions (including rare fluorspar). The frame was divided horizontally into two parts; the lower part was vertically slotted and held a horizontally-sliding slotted frame to control ventilation, while the upper half held the mortices for two vertical bars, and rebates for glass (with traces of putty). The jambs of the opening were rendered with hard brownish-white sandy cement.

The original window (Plate 24) lay near the south end of the wall, its southern jamb being 2.2 m from the external corner, and its sill 1.4 m above internal floor level. The jamb consisted of a stopped end in the original walling, abutted externally by a single thin slab laid on end, resting on a similar flat slab forming the sill, both set in Type I bedding. The north side of the sill lay just below the sill of the replacement window, and the extreme base of the northern jamb appeared to survive beneath the southern jamb of the later window, indicating a width of only 0.5 m for the original window. The window had been 0.46 m high internally and 0.7 m externally, both sill and lintels having apparently splayed outwards. The (presumed) socket for the lintels survived both externally and internally; it was 120 mm high, extended 250 mm back from the jamb externally and 460 mm internally, and was infilled by later blocking. Externally, the original sandstone masonry and the Type I bedding around it were both reddened, suggesting that a timber lintel had been burnt *in situ*.

The original window had been infilled with masonry of coursed sandstone rubble with scattered firebrick and slabs of re-used cementation chest sandstone, bedded in hard pale grey-brown crumbly lime mortar. This extended into the lintel sockets (where it was largely of brick and firebrick), and extended north and upwards to fill a break in the original wall, for the insertion of a replacement window. This was 0.88 m wide externally (1.10 m internally), and 0.96 m high to surviving wall top (here 0.2 m below the original wall head). It was splayed internally, the jambs being formed of stopped ends in the inserted masonry and the sill by a row of inclined slabs. The external sill consisted of a large sandstone slab, into which an Ordnance Survey benchmark was cut. No evidence for any lintels survived.

At the north end just below eaves level, an area of coursed rubble set in a harder variant of Type V mortar appeared to fill a socket 340 mm wide, 180 mm high, and 430 mm deep in the inner face, and was continuous with a patching to the top of the adjacent furnace buttress, over the top of the built-in beam (*Ch 5*). On the south side of this feature, an area of large coursed sandstone blocks set in hard white mortar may have been cut in, or have infilled an original feature. The external face behind both these features was occupied by an ill-defined area of disturbed masonry. Interpretation of these features was difficult due to unstable masonry and disturbance by tree roots; they can best be interpreted as representing the insertion of a beam socket, followed by its blocking when the corner of the furnace was repaired.

A small socket (120 x 110 x 110 mm) was chiselled into the internal face just north of the doorway, and 1.0 m above floor level. It was lined with brown sandy mortar, over which a pocket of charcoal dust was sealed by an infill of hard white mortar. A linear vertical scatter of black encrustations above and below this suggested the former existence of a post or partition against the wall face. A similar feature was cut into the west buttress on the south face of the furnace, and the walls between these features and the north-west corner of the building retained scattered deposits of charcoal dust (in crevices and under surviving patches of limewash).

The internal face also contained an inserted socket, 1.9 m above the north edge of the late concrete floor (*Ch 6*). From here south, the face was intermittently rendered with cement (as was the south wall and the southern part of the east wall).

The wall top contained the remnants of several timbers, forming the base of the roof structure. These were extremely rotted, and had become incorporated into the root mat of the vegetation cover of the wall top (which had penetrated deep into the sand bedding, and was holding the wall head together); only limited cleaning was possible without jeopardising the structure.

The clearest remains formed a group central to the length of the wall. Here a slot (1.35 m long x 0.15 m deep x 0.10 m high) ran along the top of the internal face, its base lying 0.3 m below eaves level. It contained traces of rotted wood, the grain running along the wall line. From its top, an east-west slot (300 mm wide x 220 mm high) ran across the wall top. This contained the remnants of a beam bedded in weathered grey-brown sandy mortar with lime flecks; the mortar did not extend into the structure of the socket, suggesting that the beam was a replacement.

A similar but less well-preserved group of features survived above the north side of the doorway; here two parallel beams or planks had occupied a slot along the wall, with traces of a cross-beam over their tops. The beams were bedded in grey-brown mortar, whereas the slot structure was bedded in Type I sand. A third group lay at the north end of the wall (over the features described above) surviving as rotted wood in the remnants of a slot, with traces of east-west wood across its top; the base of the slot had been

built up with 50 mm of small slabs, set in decomposed grey-brown mortar.

These features are interpreted as the remains of separate wall-plates, *c* 1.3 m long, along the wall face, each holding a tie-beam across the building to form the base of a roof truss. The tie-beams were a uniform 2.5 m apart. Any fourth truss would have lain over the secondary window near the south end, where the wall head did not survive. At the south end, however, a horizontal slot (260 mm deep x 180 mm high) did extend for 0.70 m along the wall head from the internal corner, with no surviving timber. This was continuous with a beam-slot extending a short distance along the face of the south gable.

At the start of recording, the south wall stood virtually intact to eaves and gable height (respectively 2.9 m and 5.5 m above floor level); its upper part was later dismantled and rebuilt, for safety reasons. The wall was of normal rubblestone construction, with more regular quoins on the corners, and some coal and brick fragments in the Type V pointing; dismantling showed the core to consist of Type I sand packed with small angular sandstone fragments, the wall being bonded by overlap of the rears of the facing stones, and by occasional non-projecting throughstones. The verge line of the gable survived in places, and was a simple angled top at *c* 42° to the horizontal.

The main feature of the wall was a blocked central opening, 1.20 m square, its sill lying 1.2 m above floor level. The sill was of timber internally; externally a stone sill may have been robbed. The lintels (internal and external) were of timbers 1.65 m long and 0.1 m high, bedded in Type I sand. The blocking consisted of sandstone masonry in hard white gravelly mortar, which extended into a partial render of the upper part. The external base of the blocking degenerated into a packing (or possibly a remnant of adjoining stratigraphy) of silty sand and coal fragments, in front of unmortared brickwork.

The lower part of the eastern verge had been rebuilt; the rebuild contained occasional brick, firebrick, burnt sandstone and red quartzitic sandstone, set in brownish-white crumbly mortar with fine gravel and coal fragments. Dismantling showed the core to consist of sandstone and firebrick in mortar, with no throughstones. The rebuild had irregular edges, and was continuous with the rebuild of the east wall (*see below*); it did not contain any beam-slot at eaves level.

Much of the inner face of the wall was covered by a thin render of off-white sandy gravelly mortar, which had been whitewashed; this extended over the blocking of the opening, but did not survive to intersect the rebuild of the eastern verge.

At the start of recording, a pair of timber partitions survived inside the building, extending north from the south wall (to which they were anchored by slots cut into the original wall and the blocking of the opening; a pair of slots cut in at a lower level may have held the base of a manger, for which the supporting brackets survived on the partitions.

In its existing form, the east wall contained only two openings, a pair of doorways (both original) in its northern part; the southern gave access to the open air, and the northern to the Eastern Building. However the southern part of the wall (now devoid of openings) had been totally rebuilt. The original wall is described first, followed by the rebuilt southern part and the features in the wall head. The wall was of normal construction, except that (despite a sheltered location) there was no Type V pointing on the exterior where this was internal to the Eastern Building. The north end abutted the south face of the furnace internally, and the southern buttress on the east wall externally, clasping the corner of the furnace in a similar manner to the walls of the Northern Building; the south face of the furnace was tusked to receive it, but no tusking was visible on the buttress.

The top of the wall beside the furnace contained a group of hard-to-interpret features, broadly matching those in the north end of the west wall (*see above*). Internally, the area consisted of sub-rectangular sandstone masonry, including several burnt blocks and a vitrified firebrick, set in fine homogenous sand (greyer than normal Type I); this occupied an area of *c* 400 x 400 mm beside the furnace, its base being level with the tops of the buttresses on the furnace south face. Externally, this was matched by a larger area of coursed masonry including firebrick and a burnt sandstone ashlar block with one vitrified face, set in hard sandy lime mortar. This occupied an area 1.08 m long by 1.07 m high on the wall face, extending to eaves level. A rectangular area of coursed rectangular blocks (including the vitrified ashlar), against the face of the furnace buttress, looked like a separate socket, though the mortar was identical and apparently continuous. This feature lay at the level of the blocked socket in the buttress (*Ch 5*), the south end of which presumably lay behind the wall facing.

The northern doorway was 0.96 m wide and 1.80 m high; its jambs consisted of plain stopped ends, overlying a sill of two slabs, containing the square sockets for a door frame. The top was lintelled by five beams (averaging 150 x 100 mm in cross-section), bedded into the wall and continuing south to the jamb of the southern doorway (where they were cut off). One of these lintels, directly over the sockets in the sill, was set 60 mm lower than the rest. A door frame survived, but did not occupy the sockets in the

sill, and was noticeably newer-looking than the lintels. It consisted of a simple frame of three timbers, each 110 x 70 mm in cross-section, with rebates in the west face. The base of its north side was held in place by a wedge in a joint of the wall. The wall between the two doorways was effectively a free-standing pillar (0.44 m long and 0.62 m thick) beneath the lintels; it was more carefully built than the rest of the wall. Above the centre of the doorway, a gap 120 mm square by 150 mm deep in the internal wall face was probably a beam socket.

The southern doorway had had a more complex history. The lower part of its jambs consisted of stopped ends 1.0 m apart in the original wall, resting on a sill of two slabs across the base of the opening, with sockets for the door-frame. However these jambs ended 1.8 m above sill level, where the lintels of the northern doorway were sawn off in the northern jamb. Above this level, the top 0.3 m of the jambs was formed in inserted masonry of small sandstone slabs with firebrick and pantile fragments, set in pale grey-brown hard lime-and-cement mortar containing coarse sand and fine gravel; this masonry was cut into the original walling on each side, and also into the rebuild on the south side. The existing lintels were built into this masonry; they were rotted and held unstable masonry, and could not be closely examined. The doorway contained a frame, 140 mm from the outer face, of three timbers 100 x 70 mm in cross-section. The bases of both sides were formed by replacement timbers scarfed diagonally to the originals, each joint being held by a square-sectioned iron rod driven into the wall beside the frame, with (on the better preserved example) a leaf-shaped head at right angles against the door face of the frame, with a central nail hole (*cf* Fig 15). An iron strip driven into the south jamb, outside the line of the existing frame, may have been strapping for an earlier frame.

The original wall formed the south jamb of the doorway to original lintel height, but was broken off to the south, surviving for a length of 1.9 m at floor level. The lower part of the internal edge of this break had been burnt *in situ*, the stones having been blackened and the bedding sand reddened. From here almost to the south end, the original wall survived only as a footing, with some evidence for the site of a doorway; this is described as part of the excavation evidence (*Ch 6*). The original wall resumed at the south end, standing to 0.4 m short of eaves level at the corner, with a jagged break to the north. This remnant preserved Type V pointing externally (no pointing survived externally near the southern doorway, where the joints were deeply weathered).

Between the original parts of the wall, the rebuild was composed of a mixture of 'normal' sandstone masonry, burnt sandstone, and firebrick (some vitrified), all roughly coursed and bedded in crumbly brownish-white mortar with coal flecks and fine gravel inclusions. Some of the sandstone retained adhering Type I brown sand on its faces, under the mortar, suggesting immediate re-use from the original wall. The base of the rebuild was unclear, the bedding of both phases having weathered into soil below modern ground level; it is thought that the footings were original, but all overlying walling rebuilt.

The internal face of the wall, from the southern doorway southwards, retained patches of rendering, of hard pale grey lime mortar with gravel, sand and lime inclusions, surviving on the faces of both the original and the rebuilt wall. The render had in turn been whitewashed. Remaining features in the wall faces were of limited importance, and are described in the Archive Report.

The wall head also contained features, relating to the roof structure. At the north end, above the group of alterations, a horizontal beam was bedded into the wall, set in Type I sand. The beam extended 1.3 m from the abutment against the furnace, and had been c 90 mm square, its base lying 200 mm below the wall head. It was severely rotted, and the overlying masonry had collapsed, so that no evidence for any overlying truss survived. Above the northern doorway, an irregularity in the wall head may have been the remnant of a similar feature; the area was too unstable for close examination.

Further south, remains of two beams, both very rotted, survived running along the head of the rebuilt wall. The northern of these (c 120 x 60 mm in cross-section) lay horizontally within the wall structure, its west side being set 130 mm back from the wall face, behind the facing blocks. On top of this, tenuous traces survived of an east-west timber (5.1 m from the face of the furnace). Both beams were set in hard grey sandy lime-mortar with coal flecks (similar to that of the heightening of the southern doorway), overlying the mortar of the rebuild. The southern beam lay midway between this and the south end; it was set 70 mm from the inner face, and bedded in the same grey mortar as the northern beam. No trace of any truss survived. Fragments of similar grey hard mortar were scattered along the top of the rebuild, perhaps derived from the eaves of the final roof.

The Eastern Building

The Eastern Building measured 4.2 x 2.8 m externally; its west side was formed by part of the east wall of the Southern Building, and the west part of its north end by the southern buttress on the east face of the furnace, both of which it abutted. Its only entrance was by the doorway from the Southern Building, described above. It had been roofed by a catslide extension of the Southern Building roof. Its walls were of similar construction to those of the Northern and Southern Buildings, including the use of Type I sand bedding.

The structure of the west wall has already been described, as part of the Southern Building. Its east face showed no trace of Type V pointing (despite the sheltered location); instead it had been pointed or rendered with a cream-white sandy lime mortar (with no coal inclusions), surviving in patches; less extensive patches also survived on the other internal wall faces of the building. It was overlain in places by a late limewash (*see below*). A socket had been cut into the wall, 0.6 m from the north end and 0.4 m above floor level; the north side was angled at c 30° to the line of the wall, aligning with a socket in the north wall (*see below*). Several nails were driven into the wall; details can be found in the Archive Report.

The north wall abutted the inclined face of the buttress of the furnace to the west. Its exterior face was quoined at the corner, and was pointed with Type V mortar, which also extended as a repointing over the north face of the buttress. The inner face had collapsed at the east end. Most of the verge had been rebuilt, but enough survived to indicate that the original line had been similar to that of the rebuild. A horizontal length of timber was built into the internal face (and bedded in Type I sand) at eaves height; the west end was 0.55 m from the internal corner, and a length of 0.27 m survived. It may have been the remains of an 'L'-shaped wall plate on the corner.

Most of the original wall top had been replaced by a rebuild of very roughly coursed sandstone masonry with burnt stone, burnt and unburnt brick and firebricks all set in Type II mortar. To the west, this cut across the butt joint between the wall and the buttress, replacing part of the face of the buttress. Towards the west, it incorporated a socket (perhaps for a purlin) across the verge of the wall, blocked with brick set in Type II mortar. Further east, and central to the length of the wall, a purlin survived, set in a socket across the (rebuilt) verge. It consisted of a rectangular beam 160 x 60 mm in cross-section, laid on edge at right angles to the pitch of the roof. It was sawn off flush with the external wall face, and rotted off above the internal face. Above this lay a rotted rafter (60 x 50 mm in section), laid along the centre of the top of the rebuild and set in Type II mortar. Across this survived traces of a single lath, 35 x 15 mm in section. Over these, remains of a pantile roof survived, consisting of a row of pantiles along the verge; most of these had slipped from their original position, and no evidence for their attachment to the roof structure survived. The pantiles approximated to rectangles of 340 x 240 mm, laid with their long axes up the roof; the northern top corners were rounded off, and the southern lower corners cut off at 45°. The roof as a whole lay at an angle of 35–40° to the horizontal.

The lower part of the wall and buttress were covered by a cement render, continuous with the late concrete floor (*Ch 6*). This was covered with limewash (which also survived in patches directly on the walls of the building). The limewash coated the sides of a square socket, cut into the face of the buttress 0.4 m above floor level; this probably matched the socket in the west wall, described above.

Unlike the other walls, the east wall had partially collapsed; its south end stood to eaves level (1.4 m above the internal floor), but the remainder was broken off, only 0.4 m above floor level towards the north end. The corners were quoined externally, and the core was composed of irregular sandstone fragments in Type I sand. No Type I pointing survived, the external face being very weathered. The final eaves survived at the south end, consisting of up to 180 mm of concrete with pantile impressions in its surface.

The wall contained the remains of a blocked window. The south jamb survived as a stopped end, 1.5 m from the external south-east corner; it was broken off just below eaves level, and its base lay 0.7 m above floor level; a soil-filled recess below this was probably the robbing-slot of a sill. The window had been blocked with coursed firebricks (and some housebricks) set in Type II mortar, the blocking being constructed as two skins each two bricks thick (bonded with headers) separated by a narrow core of mortared sandstone and tile. It survived for a length of 0.5 m from the jamb, beyond which the wall was broken off to below sill level (for a 0.9 m length).

The south wall survived virtually intact, abutting the Southern Building east wall between the doorways. The verge had been rebuilt externally; it could not be closely examined due to unstable masonry, but consisted of brick walling with a concrete bed on its top (retaining pantile impressions), at a slighlty shallower angle than the original verge.

The internal faces of both east and south walls had been broken away at their corner, and replaced by a fireplace. The base of this consisted of a smooth slab of hard fine-grained blue sandstone, laid diagonally across the corner at floor level, its front face abutting the broken ends of the primary walls. Over the rear of this, and on the same diagonal alignment, was an area of coursed brick walling with some sandstone, bedded in a lime-rich sandy mortar. This walling was heavily burnt, and had a slightly convex face. At both ends, it passed behind secondary brick walling, on the line of the building walls and meeting at right angles above the top of the diagonal walling. These walls consisted of coursed brick and firebrick (some burnt, and including brickbats), with some sandstone, set in Type II mortar; there was no evidence of *in situ* burning. The top of this brickwork was continuous with the rebuild of the southern verge.

6

THE EXCAVATIONS

Introduction

The interior of the building complex and the areas immediately outside it were excavated. The objectives of excavation were:

1   to expose any surviving internal floor surfaces and excavate other internal stratigraphy to a suitable level for consolidation and display;

2   to establish contemporary ground levels around the buildings, investigate any structures suitable for consolidation and display, and record any stratigraphy threatened during the consolidation process;

3   to expose all walls to floor or ground level for recording and structural examination.

In general, excavation was terminated at the final floor or surface levels relating to the industrial use of the site; deeper excavation was confined to areas where fragile stratigraphy would have been left vulnerable to damage during site management, and to limited 'problem-solving' excavation to establish the depth and nature of stratigraphy below the proposed consolidation levels. All these excavations were undertaken to normal archaeological standards: all layers, surfaces, features and structures were recorded on a single context system, and drawn at 1:20 where appropriate, and all finds were retained and analysed.

In addition, various service trenches were excavated or observed where needed for works and presentation purposes within the Guardianship boundary. Recording of these trenches was undertaken in context or free-text narrative form as appropriate, and finds were only retained where their stratigraphy could be reliably established.

Excavation areas within the furnace buildings are described first, followed by the areas immediately around the buildings (in anti-clockwise order from the south-west corner); description finishes with the outlying service trenches. Within each area, contexts are described as far as possible in stratigraphic order, from earliest to latest. Context numbers are used in the text only where necessary for cross-referencing.

The internal areas

The bulk of the deposits excavated within the interior of the buildings consisted of rubble from recent deterioration of the structure. In many areas, these overlay intact floors relating to the final agricultural use of the buildings (but incorporating varying amounts of earlier floors). Excavation was halted at these final floor surfaces, except where otherwise noted.

The interior of the Northern Building (Fig 9) contained a simple stratigraphy over a continuous late floor, which was not removed. The northern end of the ashpit lay within the building, and the earliest soil layers encountered lay within this, pre-dating the floor to their north; the structure and stone floor of the ashpit have been described on *Ch 5*.

The slab floor of the ashpit was stained with ash, but its hollows and interstices were covered by a thin spread of variegated yellow clay, underlying a thin layer of compacted coal ash (which also overlay the treads of the two lowest steps at the north end). To the south, this was overlain by a bank of (successively) fine coal ash (250 mm thick), coarse coal ash with partially burnt coal (300 mm thick), and burnt red sand with vitrified masses and some vitrified firebricks; this bank dipped north from beneath the north end of the firegrate of the furnace, and formed the north end of an infill of debris from the final firing, the remainder of which is described below (*Ch 6*).

The north end of the pit, over the surface of the bank, was filled by 0.9 m of dark clay-silt with yellow-orange clay lenses and some firebrick rubble. This was overlain by up to 350 mm of soft gritty coal ash

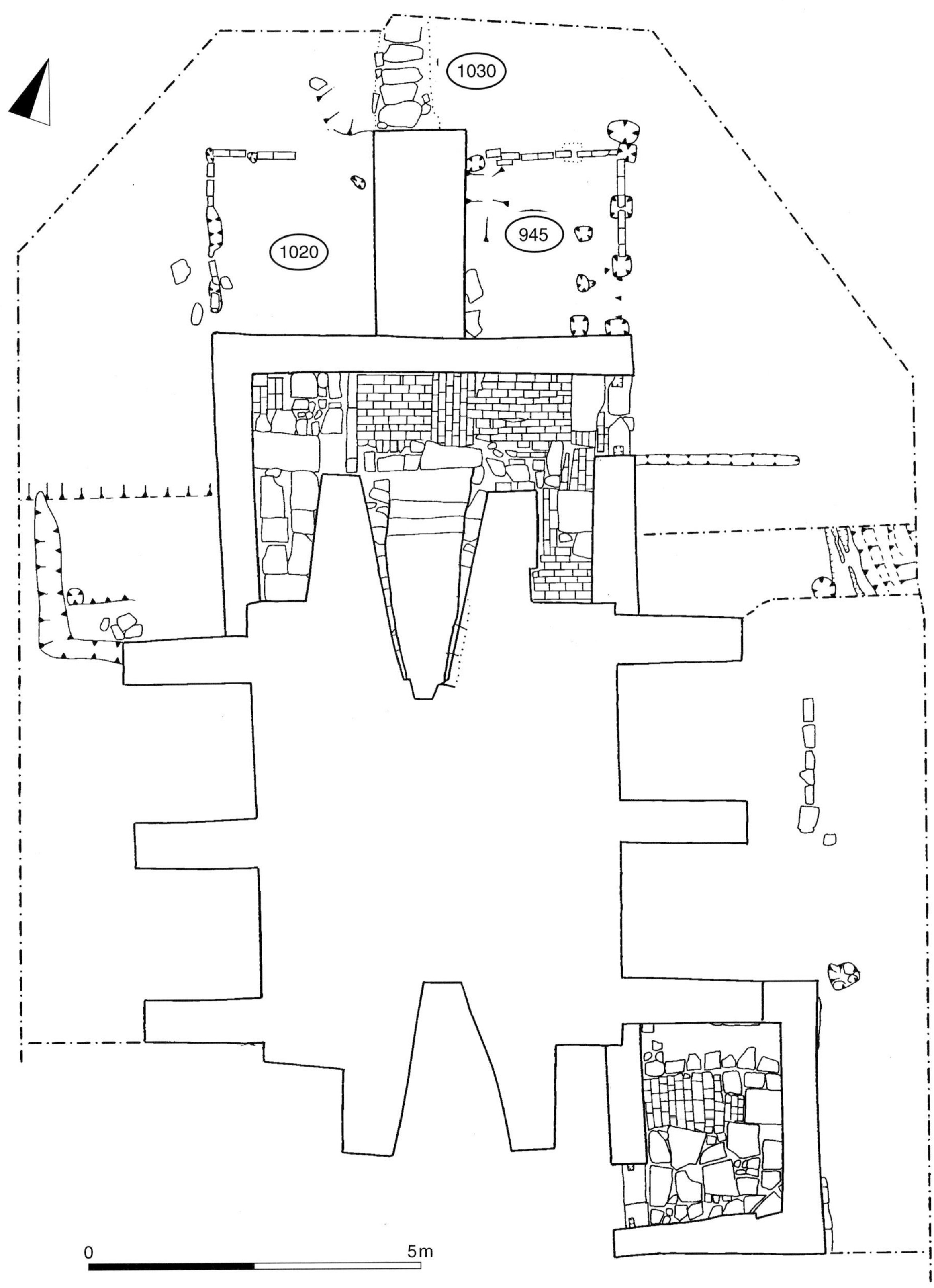

Fig 9 *The excavated areas: northern half, showing major features and structures*

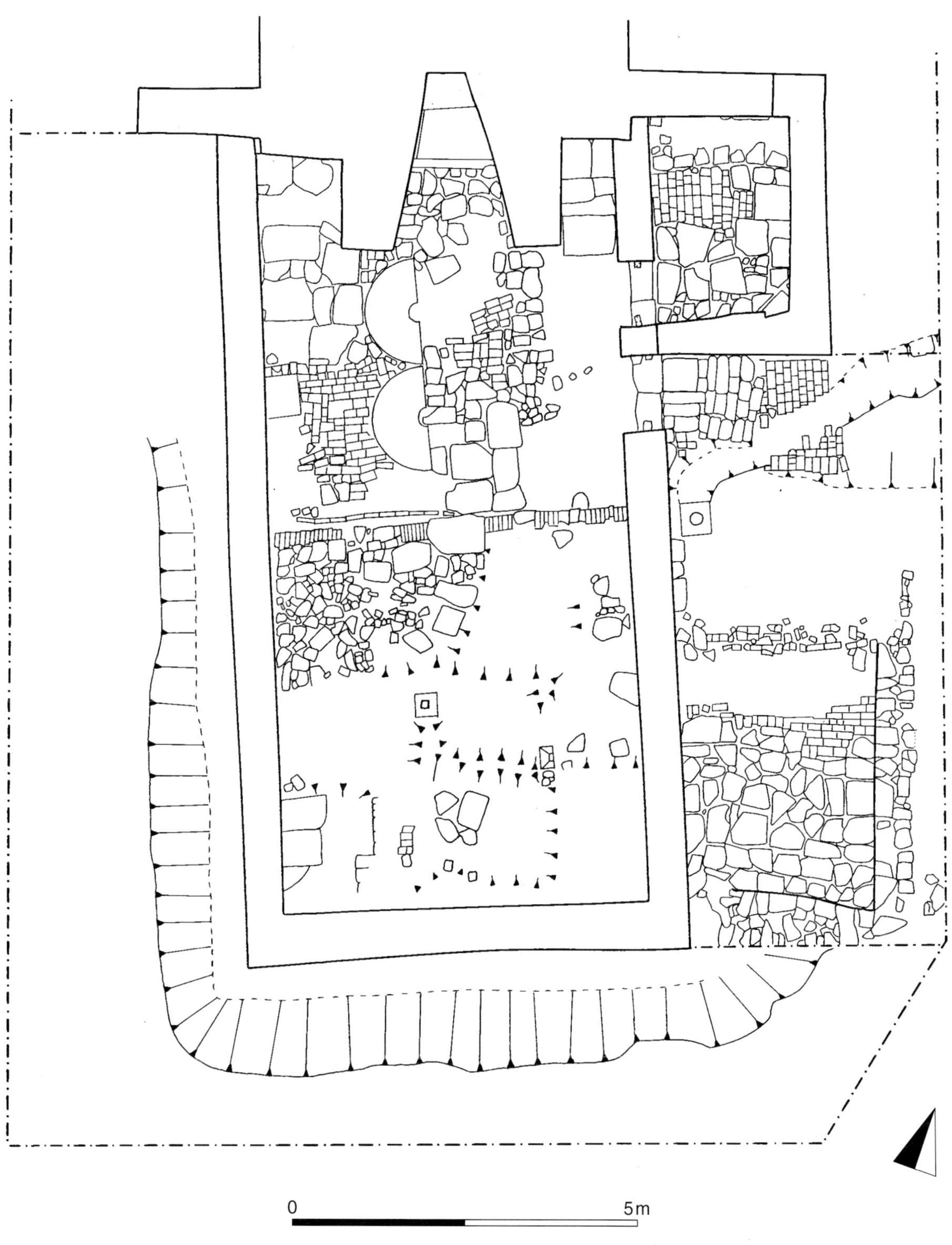

*Fig 10   The excavated areas: southern half, showing floors etc as exposed by removal of
overburden and final concrete floor*

with clay lenses, containing sandstone and pantile pieces, and quantities of patterned brown and green window glass, and dipping and lensing out to the south. To the north, it overlay the scar of the robbed top step (*Ch 5*), and continued as a layer 100 mm thick beneath the kerb of the floor of the building (which was not removed).

Most remaining infills in the ashpit could not be related stratigraphically to the floor of the Northern Building. They consisted of: grey-brown sand-silt with coal; dark sticky clay-loam with pockets of yellow clay and abundant pantile rubble; and a 'doughy' clay-loam with mortar and coal fragments. Above these, the top of the ashpit (over the sidewalls and abutting the kerb of the floor) was infilled with varied deposits of earth, rubble, and coal ash.

The floor of the building (Fig 9, Plate 25) was divided into two parts by a wall or kerb of three edge-laid sandstone blocks, running from the north-east corner of the western buttress of the furnace, north to the north wall. The two northern blocks were 160 mm wide, projected slightly above floor level, and their tops were rounded by wear. The southernmost block was wider and higher, and its arrises were unworn. The structure probably formed a threshold or the base for a partition.

To the east of this, the floor consisted of brick and firebrick (some stamped) laid on bed, with scattered sandstone flags. It abutted the sill of the doorway; to the north, between the buttresses of the furnace, it ended at a kerb of sandstone slabs, just short of the end of the ashpit and underlain by the lower infill of the pit. It is likely therefore that the whole floor post-dated the closure of the furnace.

To the west of the partition, the floor consisted of roughly-laid sandstone flags with pebble and brick packing in the joints. The flags were smaller and thicker at the north end. Opposite the end of the buttress, an east-west line of small edge-laid blocks protruded by up to 100 mm from the adjacent floor, returning south as a line of firebricks flush with the floor level. This feature may have been the base for a partition or cupboard. In the north-west corner, the stone floor had been replaced by a patch of brick and firebrick flooring, with a slightly dished surface, beside a small opening broken through the wall (*Ch 5*); it may have been the base for a trough or tank draining through the hole.

The floor was directly overlain by topsoil, consisting of 50–400 mm of dark crumbly loam with large amounts of pantile rubble and some sandstone, slag, and firebrick. Along the east side, this passed under a bank of sandstone rubble in grey loam, 0.6 m high, derived from the collapse of the east wall. The surface of the area was covered by loose rubble (with firebrick in the firegrate arch area), in a mat of soil and vegetation.

The stratigraphy of the Southern Building was more complex (Plates 26, 27). The building was terraced into the natural slope of the ground, but its internal floors still sloped up slightly to the south. Its stratigraphy divided into two areas. In the northern half, removal of recent overburden revealed a stone and brick floor of composite origin; only limited areas were investigated further. In the southern half, removal of recent overburden and a late concrete floor revealed a thin but complex stratigraphy, with fragments of floor surfaces. The earlier and more solid floor surfaces were left intact, but the remaining stratigraphy was excavated to the surface of primary make-up, since it would have been very vulnerable to damage during consolidation and display. This southern half is described first (Figs 10, 11).

The surface of natural-like deposits (Figs 10, 11), beneath the stratigraphy of floors, was exposed through much of the south half and locally in the north half; in the former area it lay well below the surface of undisturbed natural Boulder Clay outside the building. However much of this surface appeared to consist of redeposited rather than *in situ* natural; the distinction between the two was not entirely clear.

In the south-east corner, excavation revealed a deposit of firm brown clay with patches of grey soil, buff sand, red sand, and cobbles in its surface. Where cut by a later feature, it overlay a layer of dipping sandstone slabs bedded in similar clay. This gave every appearance of being *in situ* natural Boulder Clay, with a disturbed surface, and was cut by construction trenches for the walls.

The construction trenches were largely filled by the south and east walls and their footings; the face of the south wall continued to the base of exposure within the trench, but the east and west walls had footings of sandstone masonry, projecting up to 120 mm from the overlying wall face, with a flat top which in places stood as much as 120 mm above the adjacent surface of natural or primary make-up. The footing of the east wall butted off 2.8 m north of the internal south-east corner, and resumed 1.02 m further north; these ends were abutted by primary make-up, and overlain by the continuous face of the rebuild of the wall (*Ch 5*), and they may have marked the site of a doorway in the original wall. The remaining infill of the construction trenches consisted of pale brown sand (clayey in places) with sandstone pieces and some charcoal and iron-staining. These trenches were only visible in the south-east corner.

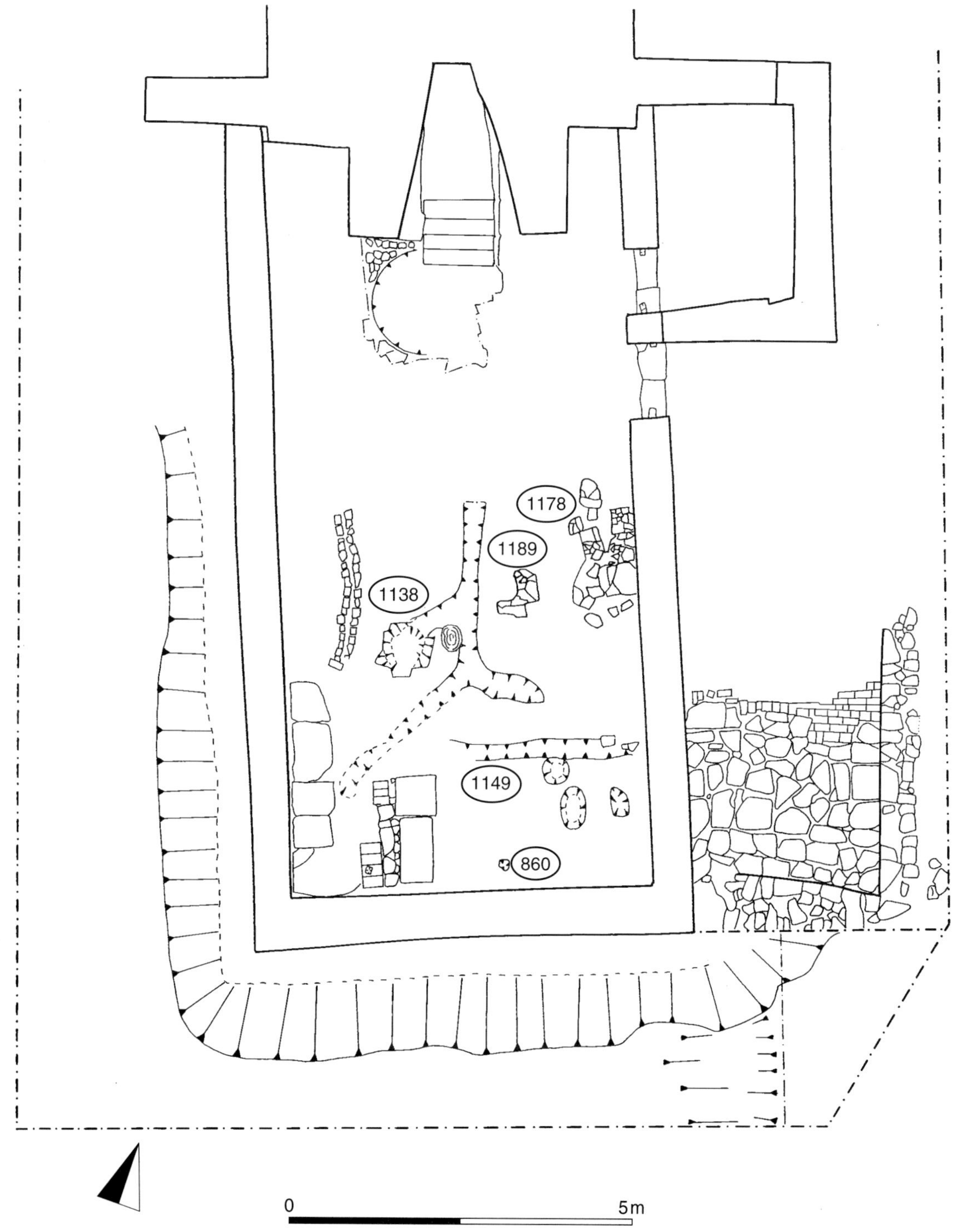

Fig 11   The excavated areas: southern half showing major earlier features

The remainder of the area, where not obscured by un-removed later floors, was occupied by more mixed deposits, all of which at first sight looked natural in origin. These varied from hard yellow-brown clayey sand with rare pebbles and sandstone pieces, to hard light grey-brown silty sand with sandstone rubble in its surface, to stiff buff clay with stones and sand patches, and very hard yellow-brown sand with pebbles and dense sandstone pieces. No consistent stratigraphy could be seen within these deposits, and (where visible) they directly abutted the building walls and footings without visible construction trenches. Observation of a service trench (dug across the interior after excavation, just north of centre) showed a thickness of 0.7 m of hard dirty brown sand with rare charcoal flecks; beneath the east wall, this abutted a vertical face of harder 'clean' yellow sandy clay, which also appeared to underlie its base. The

55

evidence therefore indicates that much of the building was built in or over a deeper pre-existing terrace into natural ground, the majority of deposits visible at the base of excavation being backfills of redeposited natural within this.

In the centre of the building, sharp vertical edges were observed between the 'natural-like' deposits, suggesting the existence of backfilled features within the primary make-up; these could not be elucidated. Two features were also observed to be stratified within or under the 'natural-like' deposits here. One of these, in the very centre of the building, consisted of a very large irregular rotted post or (more probably) tree stump (Plate 28), widening downwards and measuring 320 mm square. Its sides were abutted by hard buff sand which may have been part of the primary make-up, or backfill in a separate pit. To the west of this, the sides of a large post-pit (1138, *see below*) revealed at least 150 mm of hard homogenous sand, varying in colour from bright brown through buff to black (the black lenses smelling of hydrogen sulphide), and underlying the apparent primary make-up. It contained a single large rounded sandstone rock beneath the base of the post-pit. These features were undoubtedly stratigraphically early, their surface underlying the sequence of floor deposits; they are likely to have dated from the primary construction of the building, but this cannot be proved.

Above the surface of the early deposits just described (which was flat apart from a gentle downslope to the north) the sequence of deposits relating to the industrial use of the building was only 100–200 mm thick. However it consisted of a surprisingly intricate accumulation of thin horizontal layers, divided by later features and areas of contemporaneous erosion, and often repetitive in composition. Only limited correlation of these deposits proved possible, and many were effectively uninterpretable. Consequently only a summary description is attempted here; details can be found in the Site Archive (1988 Archive Report, 4–19). An overall division can be made between 'Early Industrial' and 'Late Industrial' features, defined respectively as pre-dating and post-dating the rebuilding of the east wall; many deposits however cannot be assigned to either sub-phase, since their stratigraphic relationship to the rebuild could not be established. The better-stratified deposits (forming the sequence along the east side of the building, from the doorway southwards) are described first, followed by the less well-stratified deposits in the west half of the interior.

In the east side of the area, several fragments of stone flooring (1189) survived, consisting of thin irregular crazed grey sandstone slabs. Most of these patches rested directly on the surface of primary make-up, one being separated by 10 mm of dark grey sand-silt with charcoal flecks, over a streak of iron-stained sand. Towards the centre, a rectangular posthole (100 x 140 mm, by 200 mm deep) was driven into the primary make-up; it had hard sides, tapering to a point base, and was infilled with firm dark sand-silt with pieces of sandstone.

The surface of primary make-up contained a distinct hollow in the area between the two lengths of footing for the original east wall. This was infilled with soft grey clay containing small coal lumps, up to 120 mm thick, which abutted the primary footings and passed under the base of the rebuilt wall. To the north-west, it passed into a bright yellow-brown silty sand, and to the south-west both these layers merged into a hard grey to brown sand-silt, containing pieces of sandy red brick.

The broken edges of early floor 1189 were abutted by a thin layer of hard pale pink coarse sand. This was overlain by up to 40 mm of hard black sand-silt and coal dust, extending over much of the east side as far north as the southern doorway; it abutted the footing of the original east wall below the base of the rebuild, so that its relationship to the rebuild is unknown. This in turn was overlain by a surviving patch of floor (1178), of thin irregular sandstone slabs (rather browner and thicker than earlier floor 1189). The floor survived over an area of 1.4 x 0.9 m along the east side; its south end contained two superimposed layers of slabs, presumably indicating a local patching.

The rebuild of the east wall rested on the original wall footings where these were present, and on the surface of the grey clay layer where they were absent, with no separate footing. Here its base was abutted by a layer of firm dark sand-silt with coal lumps, filling a slight hollow in the grey clay. Over this, two plano-convex saucers of rather friable slag adhered to the wall, having seemingly solidified *in situ*. These, and the wall, were abutted by an area of hard-surfaced brown sand-silt, containing burnt sandstone fragments and flat unburnt pieces. Much of this layer directly underlay post-industrial deposits, but to the north and west it passed under later industrial layers.

The remaining Later Industrial deposits along the east side were divided into two sequences by a late east-west slot (1149, *see below*). To the north of this, the centre of the east side was occupied by a sequence of thin layers of hard sand-silt and coal dust, incorporating a fragment of a third floor of thin irregular sandstone slabs.

The sequence to the south of the slot was more complex. Over the earlier features described above, the whole area was occupied by a layer of firm brown-yellow fine sand with grey silt laminations, up to 120 mm thick. Its surface contained two firebricks laid at right angles to the east wall, perhaps the remnant of a flimsy internal structure. This was cut by a small sub-rectangular pit (380 x 230 mm, by 80 mm deep) filled with firm fine sand with coal fragments, and intermittently lined with small sandstone pieces. A second, rather similar, feature was present at a slightly higher stratigraphic level, separated by the deposition of a thin layer of dark sand-silt with coal fragments. This in turn was overlain by a layer of stiff clayey sand with sandstone (some burnt) and burnt brick pieces (863, *see also below*).

This was overlain by the final extensive industrial layer in the area, a deposit of alternating lenses of coal waste, blue-grey sandy clay, and yellow-brown clayey sand, up to 100 mm thick in total. It occupied the whole south-east part of the interior, bounded to the north by slot 1149, and ending to the west on a line joining two post-holes. One of these measured 400 mm in diameter and 230 mm deep, with no trace of any post-pipe; the other survived as a later post-removal pit (860, *see below*), the post having probably been in use during deposition of the layer.

The slot dividing these sequences was 3.6 m long, 240 mm wide, and up to 80 mm deep; it lay 2.0 m from the south wall, and marked a slight drop in level to the north. It was steep-sided and flat-based, with an intermittent lining of flat slates; both the fills and the edges were very compacted. Its fills were varied, and contained fragments of concrete. The feature is interpreted as a beam-slot, the feature remaining at least partly open until the conversion of the building to agricultural use.

The sequence just described, occupying the east half of the interior, was cut off to the west by a late drain along the centre of the building. To the west of this, much of the interior had been truncated by a wide shallow feature of late date; the surviving features were therefore rather poorly stratified, and could not be reliably correlated to the sequence in the east half. Only the more significant features are described, from south to north.

In the central part of the south end, two features were cut directly into primary make-up. One of these appeared to be a rectangular pit filled with hard dark sand-silt, partly concealed by a later floor of iron plates (*see below*); the other survived as an arc of hard brown sand-silt around later post-removal pit 860 (*see above* and *below*), and may have been the construction pit for the same post (which, if so, was in use throughout the industrial use of the building).

Both were overlain by an extensive layer of hard dark sand-silt with varying amounts of charcoal dust, coal, and crushed red sandstone. It was overlain by up to 100 mm of firm buff fine sand with sandstone pieces and crushed brick; to the east, this was in turn overlain by a deposit of hard red fine sand mixed with crushed burnt stone, with a straight south edge 0.64 m from the south wall. Further west, primary make-up was overlain (where visible) by 40 mm of very hard dark silty sand with pebbles.

This had formed the bedding for a group of seemingly contemporaneous solid floors (which were not removed) occupying the south-west corner of the interior (the surviving edges being largely due to robbing, except where they abutted the south and west walls). The east edge of this group consisted of two flat iron plates, measuring respectively 950 x 480 x 25 mm thick and 600 x 580 mm x 50 mm thick. Its east edge was probably original, and was abutted by layer 863 in the sequence of the east side. Its north and west sides were adjoined by small areas of brick, firebrick and sandstone flooring. To the west of these (but largely separated by an area of robbing), the west wall was abutted by a strip of stone flooring 0.6 m wide, extending 2.4 m from the south end. This was composed of well laid sub-rectangular flags, 80 mm thick. The interstices were filled with hard ferruginous sand-silt, and the surface was partly covered by hard iron-concreted sand-silt containing brick and tile fragments and coal dust. The north end was bedded on a thin (10 mm) layer of soft pink coarse sand, which continued for 0.8 m to the north, resting on primary make-up.

The remainder of the west central area had been truncated to near the surface of primary make-up. The main surviving feature was a small drain running north from the north-east corner of the surviving floor, in a small trench cut into the primary make-up. The drain was walled with flat unmortared brickbats, leaving a cavity 40–60 mm wide, infilled with dark sand-silt. The capping only survived at the south end, where it consisted of small thick angular sandstone blocks, abutting the flag floor and giving way northwards to worn housebrick.

The centre of the area had also been truncated, and was also cut by a late drainage channel (*see below*). The main surviving feature was a complex pit (1138, *see also above*) (Plate 29), just east of the tree-stump bedded in primary make-up (*see above*), the upper fills of which had been largely removed by the drainage channel. The pit measured 800 x 780 mm at the surface, and was 550 mm deep. Its upper sides were sloping, with distinct sub-rectangular recesses (giving an almost cross-shaped plan 100 mm below the surface); the lower sides were vertical, and the base was not clearly distinct from the multicoloured sand under primary make-up here. The fills (insofar as they survived) were complex, with signs of recutting; they were very hard and had been crushed into the sides of the feature.

Remaining Industrial-period deposits survived locally in spines between later disturbances, or formed narrow strips abutting the edges of the stone floors in the south-west part of the interior; they were too fragmentary for interpretation.

The earliest feature demonstrably post-dating the closure of the furnace was a bifurcating drainage channel. This originated as two arms flowing from the south-west and the east to the centre of the area, and then northwards along the axis of the building (where it was also fed by a spur from a sump over pit 1138). The south-west arm contained a ceramic pipe, and there was also a short length of built brick drain; the rest of the feature formed an unlined gully, averaging 200 mm deep, with varied infills. To the north, the drain passed beneath the floor of the north end (*see below*), but was re-located beyond as a gully cutting into the partly-infilled ashpit of the furnace. The infills contained a small assemblage of late nineteenth-century (or later) pottery; stoneware jam-pot sherds may date from *c* 1900 (*Ch 7*).

The south-west arm of the drain originated in a large hollow, apparently formed by robbing of the surrounding stone, brick and iron floors. This was infilled with firebrick rubble in brown gritty sand. The surface of this formed part of an extensive trampled surface, occupying most of the southern part of the interior, of hard black silty sand with a variable coal content, passing down into various earlier deposits. It also abutted the padstones for two posts, which had projected through the overlying concrete floor (*see below*) to form the end-posts of the timber partitions visible at the start of excavation. At the centre of the south end, the surface was overlain by an area of brown-white mortar with sandstone and coal pieces, probably builders' debris from the blocking of the central opening in the south wall. Elsewhere there were small areas of laid brick and thin iron plates, probably remnants of more extensive flooring.

The northern part of the area, from the central drainage channel to the west wall, had been truncated at this time, forming a shallow steep-sided hollow cut through the earlier stratigraphy to the surface of primary make-up. This was infilled with flat sandstone and firebrick rubble, roughly laid in places (especially where it formed a rough capping for the drainage channel). Its north end was defined by a kerb, running across the building just south of the doorways, largely composed of brick and firebrick. This defined a drop of 100–150 mm to the north, beyond which late floors and recent infill directly overlay primary make-up.

Above these deposits, the whole area from the kerb to the south end was occupied by a layer of assorted rubble and debris, forming the bedding for a final concrete floor. This was 50–100 mm thick, and had been cast *in situ* as a single slab, its north edge standing proud against recent overburden, on the same line as the underlying kerb just described. Despite this, the concrete was of two well-marked compositions; some areas contained red crushed tile aggregate, whereas others contained grey angular fluorspar, galena, and other veinstuff aggregate, presumably indicating the use of jigger tailings from a lead mine. The padstones for the timber partitions had been cast into the concrete, and the partitions were marked by grooves in the concrete. The surface was channelled, and the lowest part fed into a ceramic pipe, draining out through a hole in the base of the east wall.

In the northern half of the building, the stratigraphy was very much simpler. Over much of the area, primary make-up was directly overlain by a stone and brick floor (of composite origin), which was itself directly overlain by recent overburden. Only a limited area of the floor was removed (to fully expose the ashpit); the ashpit however was fully excavated, involving the removal of its post-abandonment infills.

The only stratigraphy relating to the industrial use of the site to be removed was in the centre of the building, where the south end of the ashpit was investigated in a limited excavation through the later floors; most of the deposits investigated in fact related to a semi-circular socket immediately south of the end of the ashpit.

The earliest deposit encountered in this area was a surface of stiff yellow clayey sand with occasional sandstone pieces, dipping gently to the north and cut out by the construction pit of the ashpit. Similar material was exposed elsewhere in the northern interior, wherever the stone and brick floors had been robbed off; it appeared to be a single deposit continuous with the primary make-up of the southern interior, and (by analogy with this) is likely to have been redeposited rather than *in situ* natural.

The surface of the clayey sand rose abruptly by 40 mm to the west, along the edge of the exposed area, defining a recessed area bounded to the north by the western buttress of the furnace. The base of this area was occupied by 50 mm of firm black coal dust and coal fragments, the south end of which was not exposed. Over this, much of the recess was occupied by the semi-circular socket already mentioned; the sides of this were formed by two triangular areas of stratigraphy in the corners of the recess. In the north-west corner, this contained an area of small sandstone setts bedded in black soil, overlain by 30 mm of hard black sand-silt with coal fragments. The south-east edge of this was iron-stained, and formed the curving margin of the socket. The area immediately south-west of the socket was only partly exposed; it was largely occupied by a single sandstone slab (80 mm thick), respecting the socket and overlain by a firm black mixture of charcoal dust with sand-silt, with charcoal flecks and clay streaks. This formed the south-west margin of the socket, and passed under later floors away from this.

The socket itself (Plate 30) was accurately semi-circular in plan, measuring 1.54 m in diameter and 40 mm deep, with vertical sides and a flat base. Its sides, and the base around the circumference, were cemented hard by deposits of rust. This left little doubt that it had originally held the semi-circular iron plate, of identical size, which was recovered from a slightly higher stratigraphic level (*see below*). It was infilled by part of soil layer 012 (*see below*). To the east, the socket and adjoining layers were cut out on a line running south from the south-east corner of the buttress; this cut had re-exposed the surface of primary make-up, and to the north had truncated the original top step of the ashpit and the brick extension to this (*Ch 5*), revealing that the semi-circular socket passed over the riser of the original step, but respected the riser of the brick step. The fills in and over the cut were stratified within the sequence of infills of the ashpit. A deeper gully was cut in along the east side of the area; this was also stratified within the infilling of the ashpit.

The east edge of this gully ran south from the south-east corner of the ashpit, passing beneath later floors which were not removed. It revealed the profile of two layers, of coal dust and coal fragments under yellow-grey clayey sand with dense sandstone fragments. These abutted the edge of an area of flat sandstone slab flooring, measuring 2.5 x 0.8 m, which was continuous to the north with the top of the east wall of the ashpit. The slabs were very cracked, worn, and shattered; they were broken away to the east, south, and west, and abutted by later flooring (*see below*). Although they had remained exposed as part of the final floor of the building, they were clearly of much earlier origin, appearing to be an original feature.

The infills of the north end of the ashpit have already been described (*Ch 6*). The central part, beneath the firegrate of the furnace, was excavated in conjunction

with the south end. The floor of the ashpit was covered by a thin hard layer of coal dust, developing into a slightly thicker (50 mm) layer of hard grey sand-silt and coal dust with yellow clay patches at the south end. Beneath the firegrate, the remainder of the ashpit was infilled with 0.7 m of alternating lenses of loose white ash, vitrified ash, and coke fragments. This dipped south within the south end of the ashpit, lensing out 0.6 m short of the bottom step.

Above the ash, the southern ashpit was infilled by a sequence of quite firm earth layers, mostly consisting of yellow-brown clayey sand with varying quantities of dark earth and brick, sandstone, and tile rubble. One of these layers contained a clay pipe dateable to after 1891 (*Ch 7*). The accumulation extended to the truncated top of the steps at the south end, but had been partly cut away further north, forming a pit in the upper fill of the ashpit. The cut was continuous with the truncation of the upper step and the area west of the socket beyond it (*see above*), and also with the cutting of a gully along the east side of the area, from the south end of the ashpit southwards. The gully was obscured by later floors to the south, but continued the line of the central drainage channel in the southern half of the building; there is little doubt that it was the north end of this drain, carrying ground-water from beneath the Southern Building and utilising the partly-infilled ashpit as a sump. The fills in the gully were continuous with those in the upper part of the ashpit, consisting of varied layers of loose earth with rubble of stone, brick, firebrick, fritted sand, and concrete. The surface of the south end had been locally packed with rubble, to form a firmer floor.

At the south end, the upper fills of the ashpit passed over the broken top of the top step, filling the cut through the east side of the semi-circular socket, as a layer of rubble set in soft silty sand, 100–200 mm thick, thinning out to the south. It was overlain by 30–140 mm of black humic sand-loam (012, *see also above*); this sealed the tops of the sidewalls of the ashpit (abutting the patch of early slab floor immediately south-east of this), and continued south beneath a group of floors in the northern part of the Southern Building which (on this evidence) clearly post-dated the closure of the furnace. Similar dark earth could be seen beneath other edges of the floors, where they had been robbed out, forming a thin layer directly over the surface of primary make-up.

The floors were rather varied in form, but appeared to be contemporary to each other. They were bounded to the south by the kerb just south of the doorways (*see Ch 6*), occupying most of the northern part of the interior, with gaps due to damage and robbing. They are described from east to west. The north-east corner was floored with worn sandstone flags, 80 mm thick. The (robbed) north edge of these was abutted by a layer of hard black earth with coal fragments, dying out onto the exposed surface of primary make-up. An area of 'pillow-shaped' wear-rounded sandstone blocks survived beside the southern doorway in the east wall.

Returning to the centre of the area, the north edge of the early stone floor (*Ch 6*) was abutted by an area of thick (150 mm) sub-rectangular slabs, extending to the kerb and passing over the edge of the ashpit. The broken west edge of these slabs had been patched with cast concrete. They were adjoined to the north-west by an area of tightly-laid firebrick and housebrick flooring. Parts of both floors were removed, to fully expose the ashpit.

To the west of these floors lay two semi-circular cast-iron plates (Plate 31), laid with their straight east edges in line. Each plate measured 1.53 m in diameter, by 50 mm thick, and contained a central half-hole 230 mm in diameter, and a line of four rusted-up bolt-holes parallel to, and 160 mm from, their straight sides. The southern plate was removed; it lay exactly above the semi-circular socket described above (*Ch 6*), but was clearly separated stratigraphically by dark earth 120 mm. Its underside was worn into a wide shallow hollow, concentric with the shape of the plate. Its north-west side passed just over a patch of brick, firebrick, and sandstone floor, between the plate and the buttress of the furnace.

The area west of the southern plate was occupied by a floor of firebrick (some vitrified) with some housebrick, which had been partially robbed. To the west, it abutted the lip of an open-topped iron tank, 610 mm square and 330 mm deep, the west side of which lay beneath the blocking of the doorway in the west wall. The tank may have been fed by a (partly-robbed) drain running from the kerb across the centre of the building; it had drained via a space under the blocking of the doorway into an external drain. The northern part of the west side was occupied by a floor of rectangular sandstone flags of varying thickness.

The final floor surfaces, throughout the building, were overlain by superficial deposits of topsoil and rubble (largely of pantile from collapse of the roof).

The Eastern Building measured 2.9 x 2.0 m internally. Its stratigraphy was very simple.

The earliest features exposed were at the north end, and consisted of a south-facing strip of sandstone slabs (the face lying 0.7 m from the north wall), backed by a deposit of yellow-brown clayey sand. This is interpreted as the truncated remains of a raised bench or step. The straight south face was abutted (at the same level) by a floor of assorted sub-rectagular sandstone slabs (Plate 32), with some firebrick patching, occupying the remainder of the interior and abutting the hearthstone of the fireplace in the south-east corner (*Ch 5*). A gap showed the floor to rest, at least locally, on yellow-brown clayey sand containing sandstone and tile fragments in its surface. it was overlain by 10–80 mm of soft earth, with tile and firebrick rubble in its thicker parts, containing an 1860 penny.

This in turn was overlain by a concrete floor, the aggregate of which consisted of lead-mine jigger tailings similar to that noted in the Southern Building. The concrete extended up to 250 mm up the faces of the walls in places, forming a local rendering. The floor was cast as a single slab; it occupied the whole interior except the south end, where it had an original edge 250 mm short of the south wall, the intervening space being filled with crumbly grey sand-loam. Above these, the whole interior was occupied by up to 100 mm of dark grey-brown rubbly loam topsoil, overlain by superficial loose earth, vegetation, and rubble.

The external areas

The whole area around the furnace and its ancillary buildings was excavated, to a width of between 1.5 m and 4.5 m from the walls of the buildings. The purpose of excavation was to re-expose the ground levels corresponding to the final industrial use of the site, record stratigraphy threatened during the consolidation and landscaping process (including stratigraphy immediately below the final landscaped level, vulnerable to progressive destruction by soil processes), and undertake limited testing of deeper accumulations, both to establish the appropriate level for consolidation and to solve major problems of site development.

The external areas differed from the building interiors in that they had been exposed to weathering and soil processes for far longer, and that (in general) steady accumulations of stratigraphy had occurred through the working life of the site. Sequences were in general dominated by soil layers and earth-cut features, rather than 'solid' structures.

The individual areas, and year of excavation, are shown on Fig 2; the following description broadly follows these, but disregards area boundaries where these proved to divide natural units of stratigraphy.

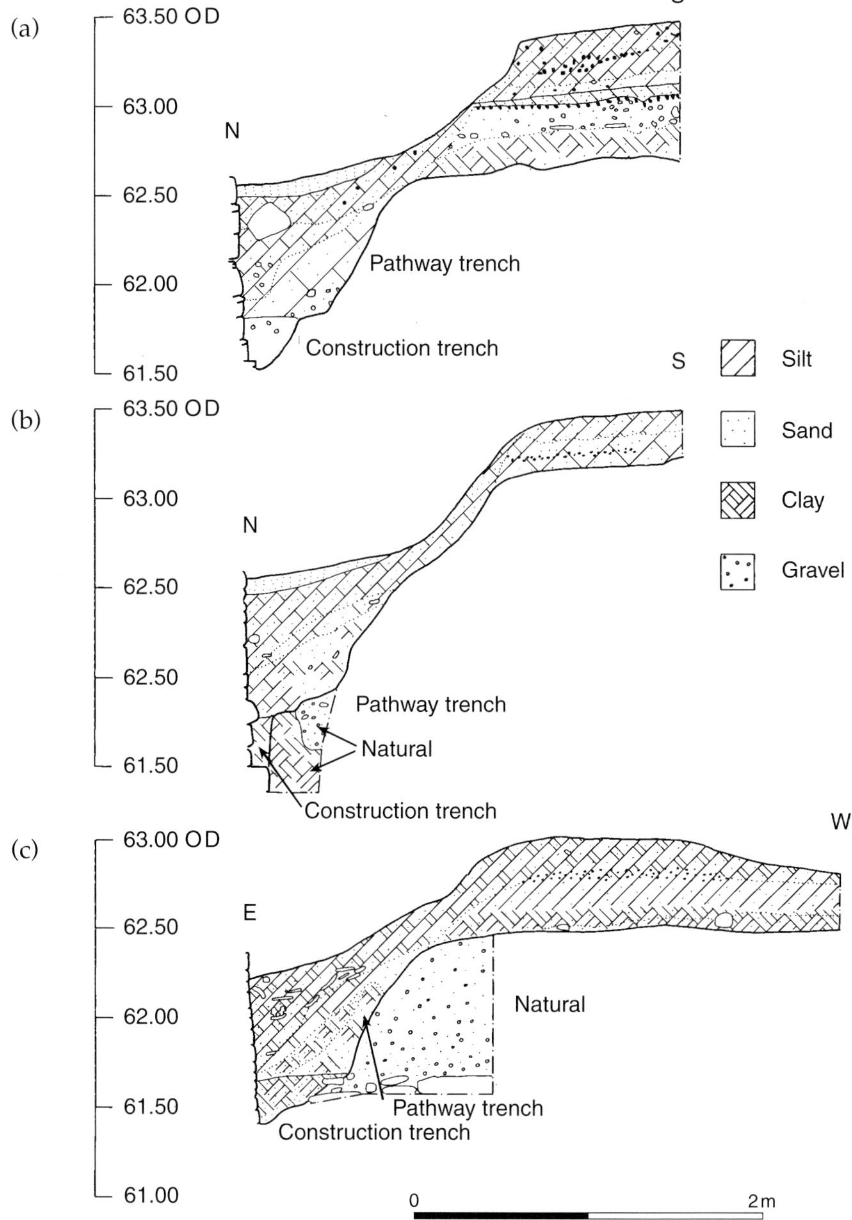

Fig 12   The southern exterior: sections of pathway trench and deposits outside it

Description proceeds anticlockwise from the south-west corner.

The southern exterior (Fig 12, Plate 33) occupied an 'L'-shaped area at the south end; its west end was separated from the south-west exterior by a baulk (removed during excavation) and its east end from the south-east exterior by the south (revetment) wall of the South-east Building (Fig 3), which extended almost to pre-excavation ground surface. Before excavation, the surface topography was dominated by a steep slope down towards the walls of the Southern Building (by up to 1 m over a distance of 1.6 m), though the ground surface beside the walls still lay *c* 1 m above the internal floors. The area outside this scarp sloped down gently to the north, and also to both east and west from the midline of the building. Excavation revealed that the steep slope reflected the partly-infilled form of a pathway, cut into natural round the west and south sides of the Southern Building and truncating a separate earlier construction trench for the walls. To the south, this had cut through a considerable stratigraphy (up to 0.9 m thick from ground level to the surface of undisturbed natural), whereas to the west of the scarp topsoil passed almost directly down into natural deposits. The area was fully excavated to the surface of natural, except for a small area at its east end (where only the latest deposits were removed).

Undisturbed natural deposits were sectioned in a test trench at the junction with the south-west area, and observed in profile in the side of the pathway trench. The earliest deposit observed in the test trench was a very hard brown clayey sand containing many cobbles and large pieces of sandstone. It graded up into loose gravelly sand, which in turn merged up into 1 m of hard yellow-brown gravelly sand with many pebbles and cobbles. These deposits continued south and east (in profile) as hard brown sand with common cobbles and patches of brown clay and coarse yellow-brown sand, merging east into stiff yellow-brown clayey sand with many sandstone pieces and some cobbles.

Above these deposits, the surface of natural was removed over the east part of the area. Here it consisted of hard brown slightly clayey sand with scattered cobbles and brick, slag, and charcoal fragments, up to 250 mm thick. This gave way westwards to firm homogenous yellow-brown slightly clayey sand containing only occasional sandstone and slag fragments. Both layers contained a thin lens of strong yellow-brown mottled silty sand, probably a weak ironpan formation. Beyond the south-west corner of the Southern Building (where overlying stratigraphy died out), this continued as 100 mm of firm yellow-brown clayey sand with rather more sandstone, charcoal, coal, and brick fragments. This became darker to the north; the surface consisted of grey-brown sand-silt, becoming browner and cleaner downwards. The whole deposit is interpreted as a soil horizon (fossil where sealed by overlying stratigraphy, but active elsewhere).

The construction trench for the Southern Building only survived beneath the base of the later pathway trench, to a depth of up to 0.5 m; it was 150–500 mm wide from the face of the wall. The footings of the walls rested on the base of the trench; they were typically 140 mm high, and projected by up to 100 mm from the overlying wall face. The south wall contained a line of four projecting throughstones, at about the level of the (later) pathway. The fill of the trench consisted of redeposited natural (with a similar range of compositions to the *in situ* natural, from which it was not always readily distinct).

The south end of the area was occupied by a stratigraphic accumulation up to 0.9 m thick, overlying natural and cut out to the north by the pathway trench. This trench had removed all stratigraphic relationships to the Southern Building and its construction trench. It should be noted that the upper edge of the pathway trench, cutting this stratigraphy, was almost certainly the product of weathering (by up to 1.0 m from its original line); some of the features cut by this edge could therefore have been cut or deposited after the original cutting of the trench. The accumulation became shallower to the west, dying out close to the south-west corner of the Southern Building.

The earliest feature in this sequence was a sub-square posthole, measuring 500 x 440 mm by 440 mm deep with steep sides and a flat base, cut into the surface of fossil soil beside the edge of the pathway trench, 1 m west of the south-east corner of the Southern Building. This was overlain by an extensive layer of small angular sandstone rubble (with some pebbles and crushed brick), bedded in firm yellow-brown silty sand which merged west into grey sand-loam. This lensed out to the west in line with the corner of the building, and was up to 200 mm thick. It was overlain by a thin (40 mm) but consistent layer of black coal dust mixed with sand-silt, containing sandstone and crushed brick fragments, itself overlain by 60 mm of hard dull red mixed sand-silt and crushed brick; both these layers lensed out 0.4 m west of the centreline of the Southern Building.

All these layers were cut out to the south-east by a hollow filled by moderately-hard brown silt with pieces of crushed brick, rounded sandstone, and coal. This thickened to the south (from 50 mm to 150 mm); its east edge was not exposed. It contained one sherd of eighteenth century Delftware, and one of potentially eighteenth century slipware (*Ch 7*). The base merged into the fossil soil; the surface was slightly dished, forming a slight hollow running to the north-east, and contained longitudinal rut-like undulations. The feature looked like an unmetalled trackway worn into the underlying deposits; its line would have respected the south-east corner of the Southern Building (assuming that this was already in existence), but was interrupted by the South-east Building.

To the west of the trackway, and 1.1 m east of the axis of the Southern Building, an irregular north-south slot was cut into the crushed brick surface; it measured 600 x 220 mm, by 300 mm deep. It was infilled with red-brown sand-silt with brick/tile and sandstone pieces, under two blocks of sandstone. A similar slot was present 0.4 m further east; the upper fill (of brown sand) was of later date, but may have resulted from robbing of an original upper fill of stones. These features may have been associated, perhaps as paired ruts. The western slot was overlain by the edge of a layer of firm brown sand-silt with charcoal flecks, up to 150 mm thick, which continued to opposite the south-west corner of the building, and appeared to be a fossil soil. The eastern part of this was overlain by a substantial deposit of angular sandstone fragments in a matrix of yellow-brown silty sand, perhaps builders' debris. Finds included the neck of a glass bottle dateable to 1730–1745 (*Ch 7*); other fragments probably from the same vessel were present in stratigraphically-adjacent contexts. It lensed out to the east, forming the west side of hollow 459 (*see also below*).

The east side of hollow 459 (over the early trackway) was formed by a sequence of layers of varying composition, all of which lensed out, or were cut out, to the west close to the projected line of the east wall of the Southern Building; the topmost layer, a varied deposit of hard red-brown sand-silt with many lumps of burnt clay, pieces of sandstone, and patches of charcoal dust, occupied the whole south-east corner of the area under late deposits, and was cut out to the north by the construction trench for the South-east Building. Only the west edge of this sequence was removed, to a section line just outside the line of the east wall of the building.

Hollow 459 was 1.8 m wide and 0.25 m deep, running north-south and dipping to the north. It was infilled with firm brown sand, very similar to the Type I bedding material. It contained several sherds of potentially eighteenth century red-wares (unfortunately the dating of this assemblage is ambiguous, *(see Ch 7)*). It was overlain by a layer of hard silty sand with some brick/tile, coal, and sandstone fragments, up to 150 mm thick near the centre of the area. This was overlain by 100 mm of quite hard dark grey sand-silt with coal fragments, which extended most of the length of the area. It was cut by a single posthole, measuring 300 mm square by 90 mm deep, and lying opposite the east side of the blocked window in the centre of the Southern Building wall. It formed the final member of the sequence to the south of the pathway trench.

The pathway trench was defined by a steep south and west side, parallel to the walls of the Southern Building, and a flat base abutting the walls; this base was consistently 0.4 – 0.5 m wide, and sloped down very gently to the west and north. The trench was 1.4 m deep beside the south wall, becoming shallower to the north due to the slope of the natural ground surface. The side of the trench was convex in profile, sloping back from almost vertical at the base to a gentle slope at the lip; this shape was presumably due to weathering, and the top edge may have weathered back by up to 1 m from its original line.

The base of the trench was occupied by a thin layer of hard trampled earth (except to the east of the window in the gable wall, where it faded out). Above this, the lower fills were dominated by redeposited natural from the side of the trench, forming a (presumably rapid) accumulation up to 600 mm thick, and becoming thinner and more humic towards the walls of the building. Beside the west wall the deposit was hard, suggesting continued trampling during deposition. These fills graded upwards into darker sand-silts and loams, the top of which contained large amounts of pantile rubble, and window glass beneath the secondary window in the west wall.

The south-east corner of the Southern Building was abutted by the south wall of the South-east Building, a north-facing revetment in a construction trench through the deposits to the south. Unfortunately, a large tree had grown from the junction of the walls; its stump and roots were removed with difficulty during the excavation, but they had severely disturbed the stratigraphy of the junction between the wall and its construction trench, and the pathway trench.

The construction trench was partially exposed in profile at its junction with the pathway trench, and its surface was visible in plan (directly beneath topsoil, the area having probably suffered some erosion) from here to the east edge of excavation. It consisted of a sloping cut through the stratigraphy to its south, largely infilled by quartzitic sandstone rubble (of blocks up to 0.5 m long) forming the core of the wall. The cut appeared to be continuous with that of the pathway trench (which returned slightly north, forming a 'Z'-shaped plan overall), and the rubble core continued south to infill the lower part of the pathway trench beside the corner of the Southern Building, over a few centimetres of rapid silting. This rubble contained lumps of fritted sand, and one large sherd of nineteenth-twentieth century white stoneware (sealed beneath a block, and below the level of severe root disturbance). The upper fill of the construction trench (of dark sand-silt with pantile and coal fragments) also appeared to be continuous with that of the pathway trench, but was very disturbed. Despite the disturbed conditions, the evidence is considered to be reasonably conclusive that the construction of the South-east Building was contemporaneous with the digging of the pathway trench.

The area to the west of the west side of the pathway trench, and of the stratigraphy at the south end described above, contained very few features. In the south-west corner, a hollow ran north-west to south-east across the surface of natural, respecting the corner of the pathway trench. The hollow was 0.7 m wide and up to 0.06 m deep. Its base was infilled with hard sand-silt with many pieces of sandstone and slag, and some burnt coal, brick, and tile, all crushed into the surface of natural. This was overlain by a lens of dark sand-silt and charcoal along the north-east side, over which the feature was infilled with dark rubbly sand-loam. The feature is interpreted as a pathway, of uncertain date. Elsewhere, scattered deposits of charcoal were present on the surface of natural, at the base of a soil horizon of grey-brown sand-silt. Two of the three sherds from this deposit can be identified as of eighteenth century date, but the context was poorly sealed (*Ch 7*).

The topsoil consisted of firm grey loam, containing pantile and sandstone rubble beside the Southern Building walls. It was up to 550 mm thick beside the south wall, and 200 mm in the areas outside the pathway trench. However it was only 20 mm thick over the top of the scarp (where active erosion was probably continuing), and was also very thin at the extreme east end (where a more marked slope to the north-east was also eroding).

The south-east exterior lay to the east of the Southern Building, and to the south of the Eastern Building (some contexts to the east of the Eastern Building are also described here, since they formed the north end of a suite of deposits within the south-east exterior area). Before excavation the area was overgrown and soil-covered, but a distinct platform was visible beside the Southern Building, terraced into the rising ground to its east and south. This proved to mark the site of the Southeast Building; the area to its north contained few features of interest, having formed the access to the doorway into the Southern Building.

The earliest material encountered was a surface of yellow-brown clayey sand with sandstone fragments, exposed locally in the northern part of the area, and in the edge of a late pipe trench here (where it contained a large rounded boulder just below its surface). It was probably the weathered surface of natural Boulder Clay, rather than of the redeposited natural observed further north (*see below*).

On the east side, the clayey sand was overlain by a surface of hard black sand-silt with coal dust. This passed westwards under an extensive deposit of pale red-brown sand-silt containing tile and sandstone pieces, which was not removed. To the north, this passed under the apparent north wall of the South-east Building; beyond this it was replaced by a hard layer of angular firebrick chippings bedded in bright red sand, the colour appearing to be produced by staining. This abutted the rebuild of the Southern Building wall to the west; it passed under the walls of the Southeast Building to the north and east, and under the floor to the south.

The south and east walls of the South-east Building (Plate 34) consisted of drystone revetments, broken off upwards. The south wall was faced with large (up to 600 x 300 x 300 mm) blocks of distinctive quartzitic sandstone, stained red on the face, and backed by a rubble core filling a construction trench (*Ch 6*). It stood to a maximum of 0.9 m high. It abutted the (original) corner of the Southern Building, and passed off the excavation to the east. This end was abutted by the east wall, which was composed of smaller red-stained sandstone (some quartzitic). Its east side was

rough, and appeared to be abutted (along the edge of excavation) by mottled red-brown sand-silt and other soil deposits. It was totally robbed out to the north, beside the end of the apparent north wall, but its line was continued for 1.2 m beyond this by a probable robber trench (infilled with topsoil), with a butt end cutting the edge of later metalling (*see below*).

The surviving evidence for a north wall consisted of a strip of brick and firebrick laid on bed, 400 mm wide and standing only 100 mm high, running from the robbed end of the east wall to the Southern Building, and slightly terraced into the underlying red bedding. An area of flat sandstone at its west end may have marked the site of a doorway.

The floor of the building consisted largely of thick irregular sandstone slabs, with some brick patching. It gave way northwards to a continuation of firebrick, resting on a thin layer of soft dark earth over the red sand bedding, and broken off short of the north wall. The whole floor sloped down gently to the north. The building measured 3.6 x 2.8 m internally.

Two successive areas of paving survived outside the doorway to the Southern Building. The earlier of these consisted of firebrick, resting on the yellow clayey sand and the red-brown bedding over this; it survived over an area of 1.5 x 1.0 m beside the eastern part of the Eastern Building south wall. Its (robbed) west edge was overlain by the east edge of the second paving, of rounded rectangular ('pillow-shaped') sandstone blocks averaging 250–550 x 150 mm by 100 mm thick, and occupying an area of 1.45 x 1.2 m beside the sill of the doorway.

Just south of the doorway, the drainpipe from the agricultural concrete floor in the Southern Building (*Ch 6*) fed through a hole in the wall, and into a piped drain in a trench running ENE across the area, cutting the earlier stratigraphy. This trench was overlain by two successive hard black surfaces, of whinstone, coal and coke fragments in a matrix of gritty sand; these had formed metallings for a track or turning area. This sequence was up to 0.3 m deep, and widened eastwards from the stone paving by the door (which it abutted).

The topsoil of the area consisted of 100–400 mm of clay-loam with some sandstone rubble, and common pantile rubble in its top. It was thin and dark over the late metallings just described and over the western part of the South-east Building floor, but much thicker beside the walls of the South-east Building, and red-brown in the area north of this, where it overlay the red bedding layers. It was overlain by mounds of loose rubble, from recent collapses of the building walls.

The east centre area was bounded by section lines extending east from (respectively) the south-east corner of the Eastern Building and the north-east corner of the northern buttresss of the furnace. The southern part of this area, beside the Eastern Building, was very simple. It was excavated onto an undulating surface of mottled yellow-brown sandy clay, merging down into clay-silt and yellow clay. This was cut to the west by an apparent construction trench for the Eastern Building wall, 0.35 m wide. This was infilled with yellow-brown silty clay with rare gravel and mortar flecks; the distinction from the material to its east was only apparent after weathering. Two postholes were cut through the trench. One (beside the north-east corner of the building) was rounded in profile and measured 500 x 340 mm by 150 mm deep; it contained three packing stones, seemingly for a post *c* 120 mm square. The other (almost central to the length of the wall) measured 140 x 90 mm, by 200 mm deep. Elsewhere, patches of the late metalling noted in the south-east exterior survived, resting directly on the surface of the sandy clay, beneath modern topsoil.

The remainder of the area, outside the furnace itself, was rather more complex. Most of the stratigraphy formed banks against the face of the furnace, divided into two by the central buttress of the furnace; these sub-areas are referred to as the southern and northern 'bays', the latter containing the more complex stratigraphy. Excavation was terminated within the lower part of the stratigraphy, at flat surfaces below the planned conservation level.

In the south-east part of the area, excavation terminated at the surface of yellow-brown sandy clay; this appeared similar to that east of the Eastern Building, but abutted the north wall of the building, with no exposed construction trench. It was overlain to the west by an area of purple-grey sand-silt and ash with coal fragments, beyond which the sandy clay rose into a north-south ridge, 50 mm wide. This was separated from the wall of the furnace by a 250 mm strip of dark grey sand-loam containing pantile and firebrick rubble.

Above these deposits, which were not removed, lay an 80 mm thickness of dark-brown clay-loam, merging into modern topsoil to the east. Closer to the furnace, it was overlain by 200 mm of very dark humic loam, containing some firebrick and sandstone (including curved pieces derived from the top of the furnace cone), and large amounts of pantile rubble at the top of the layer. This was overlain by up to 300 mm of assorted sandstone rubble in dark earth, forming the pre-excavation ground surface.

The northern bay was excavated to a flat surface composed of several layers; it was bounded to the east by a north-south kerb, beyond which the surface dipped gently to the east. The earliest exposed deposit consisted of rather dirty yellow-brown sandy clay, exposed along the east side. This can be confidently equated with the redeposited sandy clay exposed in the main test trench north of the buttress (*Ch* 6). It was overlain to the west by firm grey-brown clay-loam with dense sandstone pieces, abutting the east side of the kerb. The kerb was composed of flat sandstone blocks, with traces of a brick westward return at its south end; it was 1.95 m long and 0.2 m wide, and lay 0.9 m outside the ends of the buttresses. Its top contained two small square cut-in sockets. A single sub-square posthole (poorly stratified) lay beside its south-east corner. To the north, a stone slab east of the northern buttress, resting on an area of buff-coloured clayey sand, may have been a return of the kerb.

The kerb was abutted to the west by an area of slate-grey sticky clay; this was shown by a small test trench to pass down into grey-white clay, lensing out towards the furnace over the rising surface of yellow-brown sandy clay (presumably redeposited natural). It appeared to have levelled-up the area between the kerb and the furnace, perhaps to form the floor of a shed over the area. it was partly overlain by a thin layer of dark sand-silt with many mortar fragments.

The overlying deposits, which were removed, formed a bank against the face of the furnace, lensing out downwards close to the outer ends of the buttresses. The earliest feature was a pair of shallow ruts, averaging 80 mm wide, infilled with coarse buff sand. These pointed towards the loading hole in the face of the furnace (*Ch* 5). They were overlain by a layer of dark sand-loam with patches of ash and pale red sand, and some charcoal lumps; the layer contained a small 'raft' of laid firebrick below the loading hole. It was overlain by up to 0.15 m of dark grey mortary sand-loam; this again contained an area of laid firebrick (with some housebrick and sandstone) beneath the loading hole, adjoined to the north by an iron plate with hinges on one end (probably the door from the entry into the cone of the furnace). This layer in turn was overlain by a topsoil of soft dark loam, with some loose rubble on its surface. To the east, all these layers merged or lensed out into a dark clay-loam topsoil, separated from the early surface only by patches of black earth with coke fragments.

The north-east exterior area extended from the north-east corner of the furnace to a line north from the northern buttress on the Northern Building. Before excavation the ground surface sloped gently to the east, and more markedly to the north. Excavation revealed that this slope reflected the edge of a platform of redeposited natural around the building, immediately post-dating its construction. The outer edge of this platform (close to the edge of excavation) was overlain by a deep sequence of trackway metalling deposits, which interdigitated towards the building with a shallower and less compacted sequence, including a building (945) occupying the re-entrant between the Northern Building and the northern buttress. Over most of the area, excavation terminated at a stratigraphic horizon close to the construction of Building 945, within the lower part of the sequence of trackway and occupation deposits.

The earliest deposits were only exposed in a trench (Plates 35–37), 1.0–1.2 m wide, running east from the junction of the furnace with the Northern Building. The base of this trench was excavated 0.3 m into a deposit of stiff yellow-brown clayey sand with occasional rocks (up to 350 x 200 mm in size). This passed beneath the base of the building walls, and is interpreted as *in situ* natural Boulder Clay. Its surface dipped gently to the east, and lay *c* 1.2 m below modern ground level. It was overlain by 100–300 mm of brown clay-loam containing small amounts of charcoal, burnt sand, coal (some burnt) pebbles, and sandstone. This dipped to the east-north-east, with a hummocky surface. It is interpreted as a fossil soil.

The base of the buttress on the furnace consisted of a plinth of sandstone masonry, its surface lying just below modern ground level. This had been trench-built through the fossil soil and underlying natural, its face being encrusted with mortar which adhered to the soil (whereas the face above this level was clean). The east wall of the Northern Building, however, had been built in a construction trench 0.32 m deep, cut through the fossil soil and natural; this butted just short of the furnace buttress, leaving a thin skin of undisturbed fossil soil against the buttress (proving that the buttress had been built before the construction trench was dug). The base of the wall was formed by two courses of unmortared sandstone, each projecting 100 mm from the course above, the basal course filling the base of the trench. Above this the trench was infilled with soft mid-brown clay-loam containing some sandstone and burnt coal pieces.

The fossil soil and construction trench were overlain by two layers of redeposited natural. The lower of these consisted of 400 mm of stiff yellow-brown mottled clay with occasional sandstone pieces, and lenses of yellow sandy clay and of browner silty earth. This was overlain by 200–350 mm of hard yellow-brown clayey sand with small sandstone pieces and coal lumps, and some red sandstone fragments. Both layers abutted the faces of the buttress

and building wall. The upper surface dipped convexly to the east (becoming steep at the east end of the trench); its eastern part was corrugated by a series of ruts running north north-west, while the west side was smooth.

This surface was overlain by a more varied stratigraphy. Two hollows in the surface were filled with hard red sand and crushed red (burnt) sandstone pieces; one was overlain by more yellow-brown clayey sand. The other, close to the Northern Building, was overlain by a strip of dark sand-loam with brick and stone fragments alongside the building wall, under a wider (1.2 m) strip of stiff grey-white clayey sand with charcoal flecks and stone and brick pieces; each layer was up to 50 mm thick. They were overlain by a more extensive trampled surface of hard yellow-brown silty sand with sandstone pieces. This in turn merged into the base of a widespread layer of hard coal dust (915), passing east into a more mixed layer of black sand-silt with coal dust and coal fragments, and patches of yellow clayey sand. This layer was widely identifiable in the surface beyond the trial trench, forming a valuable stratigraphic marker.

To the north of the trial trench, a strip of yellow-grey clayey sand was exposed beside the Northern Building; it was continuous with the uppermost layer of redeposited natural in the trial trench. Its surface dipped east under a layer of pink sand with fritted sand lumps, which dipped east under layer 915 (which was not removed); it could be seen (mainly by exposure in the edges of later features) to continue north to at least the north-east corner of Building 945 (whose construction trench cut it out to the west).

The only other area where surface 915 was removed was the interior of Building 945. Here excavation was terminated at a surface of stiff yellow-brown sandy clay, dipping convexly to the north and north-east, with one rut visible in its surface in the north-east corner of the area. It abutted the northern buttress and the north wall of the Northern Building, and was presumably the continuation of the main layer of redeposited natural in the trial trench.

The earliest feature cut into this sandy clay was a posthole (not excavated) on the east side, measuring 280 x 260 mm and filled with hard grey-brown sand-loam with patches of clay and at least one piece of brick. It was sealed by a layer of soft dark sand-loam with charcoal dust or soot, which lensed out to the west and passed under the wall of Building 945 to the east. This in turn was cut by a second posthole, partly over the first, measuring 200 x 140 mm, by 120 mm deep, and filled with grey-brown sand-silt; this probably represented replacement or removal of the first post. A third posthole, to the north, was cut directly into the sandy clay, measuring 250 mm square by 220 mm deep; its fill was similar. A further posthole, to the south beside the wall of the Northern Building, measured 320 x 290 mm and contained a post-impression 200 x 160 mm by 220 mm deep; its fill was similar to that of the first posthole, and it was cut into sandy clay. These features formed a row of three posts (one replaced or removed), 0.6 m apart; there was no westward return, or northward continuation at the same spacing, but any eastward return would not have been exposed.

Further west, the surface of sandy clay was overlain by a thin layer of light grey sand-silt with mortar and sandstone fragments, under a pair of sandstone slabs (both 60 mm thick), which survived in a slight hollow. These were overlain by 50 mm of grey-brown sand-silt, with a strip of quite hard coal-rich material running north-west across its surface, cut by the construction trench of Building 945. This layer could be traced across the unexcavated surface of the rest of the area, to be continuous ultimately with layer 915. The small pottery assemblage included mid eighteenth century Staffordshire 'tavern ware' (*Ch 7*).

Over the remainder of the area, excavation terminated at the surface of layer 915 and its equivalents; this lensed out to the south-west near the east side of the Northern Building, but occupied the rest of the area. The surface passed beneath the north-east corner of Building 945, becoming less hard to the south-west. Its north-eastern parts varied in composition, some areas containing slag, clinker, and fritted sand. Near the edge of excavation it could be seen (in the edges of later features) to overlie a deeper sequence of trackway metallings, whereas towards the buildings it closely overlay the surface of redeposited natural.

Much of the walling of Building 945 (Plates 38 – 41) rested directly on surface 915; the north-east corner was separated by a local deposit of stiff yellow-brown sandy clay (presumably inserted for levelling-up, over the dipping surface of 915), whereas the west and south ends of the walls were built in shallow construction trenches through the rising surface. The building had abutted the Northern Building and the northern buttress to the south and west. Its north and east walls consisted of single thicknesses of stretcher-laid handmade bricks, surviving up to 180 mm (three courses) high; the mortar bedding had largely weathered out. The east wall ended 1.1 m short of the corner of the Northern Building, but its construction trench (with no visible separate robber trench) continued for a further 0.2 m.

Both walls were interrupted, at intervals of *c* 0.7 m, by gaps averaging 80 mm wide. These gaps overlay a set of postholes, varying somewhat in size, shape and fill (Plates 40, 41); the majority were 200 – 300 mm square, and 150 – 440 mm deep (where excavated). The fills varied from yellow-brown clayey sand to dark sand-silt; in some cases a separate post-pipe, *c* 80 mm square with a dark fill, could be detected within the larger post-pit (and directly beneath the gaps in the walling), and in one case the post-pit was cut by the construction trench for the wall. A similar posthole lay beside the corner of the Northern Building, on the projected line of the east wall. A posthole just outside the north-east corner of the building may also have been associated with a late stage of its use; this measured 500 x 360 mm by 230 mm deep. It was filled with grey sand-silt with clay lumps; this was continuous with an overlying trackway layer, and also sealed the (post-removal) posthole on the corner of the building.

Within the building, the south-east corner was occupied by an area of homogenous coarse yellow sand; this was bounded to the east by a cut through surface 915, continuing the line of the east wall across the putative doorway. This was overlain by 40 mm of firm grey-white clayey sand with charcoal flecks, fading out to the north, and itself overlain by a layer of soft grey-yellow coarse sand in the corner of the building. The sand infilled a sharp cut 80 mm deep, aligning with the centre of the east wall, and occupying most of the gap between this and the Northern Building, with a slight westward return at the south end. This may well have been the impression of a robbed threshold for the doorway.

The grey-white clayey sand appeared to continue to the west as a stiff pale grey clayey sand, occupying much of the west of the interior. This respected the north wall except for one lobe (perhaps redeposited) which crossed its broken top. Beneath it, the construction trench for the wall, and hollows in the underlying surface, were infilled with brown sand-silt, presumably an internal levelling-up.

The building as a whole was virtually square, measuring 2.6 x 2.4 m, and the gap in its east wall presumably indicated the site of a doorway. The evidence indicates a timber-frame construction, with brick infilling. The faces of the Northern Building and northern buttress preserved no evidence of the height or roof-form of the structure.

The sequence of layers overlying Building 945 can best be described here, since it was poorly tied to the stratigraphy of the rest of the area. In much of the interior, the grey-white clayey sand was overlain by yellow-grey silty sand, fading into a thin grey soil. This contained a distinct patch of dark sand-loam with charcoal lumps in the centre of the building, perhaps a bonfire site. It was overlain by a shallow rounded hollow, curving from west to south-east across the area, and infilled with stiff dark silty sand with brick/tile and coal fragments. The fill was crushed into underlying deposits, and the building walls dipped under the hollow. The fill was overlain by trackway layer 973 (*see below*), and by 100 mm of dark sand-loam with building debris; this contained intermittent lines of edge-laid sandstone slabs (60 mm wide and 100 mm high) beside the Northern Building and northern buttress. It was overlain by a short sequence of sand-silt and sand-loam deposits, under topsoil.

Over the remainder of the area, surface 915 and its continuations were overlain by various layers and features; most of these could not be related directly to Building 945, but must have been contemporaneous or slightly later. In general the sequence consisted of thin layers of earth and clayey sand to the south-west, merging into and interdigitating with a thicker and much harder sequence of trackway metallings to the north and east. Individual layers are not described here (for details see second Archive Report), only the more important features being mentioned.

To the north of the northern buttress, the lowest layers of trackway metalling were cut by the construction trench for a culvert, running north from beneath the buttress; this is described as part of the north-west exterior area. The later parts of the trackway sequence continued above the culvert, into the north-west exterior. The trackway consisted of a sequence of hard deposits, of clinker, red sand, slag and coal/coke fragments, with some lenses of sand and redeposited natural. Individual layers and surfaces were not persistent, but the deposit as a whole dipped and thickened to the north and east.

Further south, surface 915 was cut by an east-west slot or gully, extending 2.34 m from the south side of the doorway to the Northern Building. This was 150 mm wide, and 40 – 120 mm deep, and was infilled with hard light grey clayey sand. The form suggested that it represented a timber fence or wall, though it could have been a drainage gully. Its butt adjoined a tapered posthole (60 mm square and 70 mm deep); this was probably a post-removal hole within a larger and earlier post-pit, since its sides did not expose the redeposited natural through which the adjacent slot was cut.

In the south end of the area, the faces of the Northern Building and the buttress of the furnace were adjoined by intermittent lines of edge-laid sandstone slabs, similar to those noted over the site of Building 945. The purpose of these stones is not understood.

These deposits were overlain by a further extensive trackway surface (973), which also passed over the corner of Building 945. This was quite varied in composition, and extended to the south-east of the trackway itself, as a slightly-compacted grey sand-silt; within the trackway it contained varied lenses of hard sand-silt, crushed sandstone, brick/tile fragments, sand, slag, clinker, and redeposited yellow clayey sand. This was overlain by less-extensive trackway deposits, and by a rough path of firebrick and crazed sandstone slabs leading towards the doorway of the Northern Building.

These layers were overlain by a further surface of black sand-silt and coal dust, occupying most of the area and passing to the north and east under a final layer of trackway metalling, consisting of varied lenses of clinker, ash, and coal dust, with some brick and slag. This lay directly beneath topsoil, of soft dark brown clay-loam with some stone and pantile rubble to the south-west, and of dark soft loam over the trackway. Beside the building walls, the topsoil was overlain by scattered mounds of overgrown rubble; beside the north (gable) wall of the Northern Building, these contained large assorted sandstone ashlar blocks (mostly triangular in shape), perhaps collapsed from the verge of the gable.

The sequence in the north-west exterior was in general similar to that of the north-east exterior just described, though the stratigraphy was thinner (especially to the south). The north and north-west of the area was dominated by a sequence of trackway deposits, and a timber and brick structure (Building 1020) occupied the re-entrant between the northern buttress and the Northern Building. The south end of the area, to the west of the Northern Building, contained traces of a second timber building. The ground surface, and the stratigraphy, dipped to the north.

The earliest deposits exposed consisted of rather varied 'natural-like' material. To the west of the Northern Building, it consisted of yellow-brown clay-loam, merging northwards (and probably upwards) into hard yellow-brown clayey sand. This continued beneath Building 1020 as a stiff yellow-brown clayey sand with rare charcoal flecks (overlain by 50 mm of slightly greyer sand-loam with dense sandstone pieces), seen in the edges of later features to be *c* 150 mm thick, over a brown loam, which overlay more yellow clayey sand. It is suspected that this latter was *in situ* natural, the loams being a fossil soil and the overlying clayey sands the continuation of the redeposited natural whose existence was demonstrated in the north-east exterior. However the distinction between redeposited and *in situ* natural was not entirely clear without further excavation.

Wherever visible, the loam and overlying clayey sand abutted the faces of the building walls. The loam contained a single sherd of potentially eighteenth century slipware (*Ch 7*).

The sequence immediately above the clayey sands was only exposed to the north of Building 1020 and the northern buttress, and to the west of the Northern Building (since the layers concerned lensed out beneath Building 1020, and were not excavated to its west).

In the former area, the redeposited natural was overlain by a surface of hard yellow-grey sand-silt, churned a series of ruts running west north-west. The ruts were up to 200 mm wide and 80 mm deep; two were rounded, and a third flat-based. To the north of the buttress, this surface was cut by the west side of a trench, filled with soft black sand and coke fragments, with a deposit of hard grey-white powdery sandy clay at its surface. This feature was parallel to and 0.4 m outside the (later) construction trench for culvert 1030 (*see below*), and is interpreted as the trench for an earlier culvert on the same line; similar infill was visible in the east edge of the later trench. It was overlain by 60 mm of hard dark silty sand with patches of red sand and sandstone and coal fragments (1032), forming the basal exposed metalling of the trackway and lensing out to the south.

This in turn was cut by a construction trench, containing a stone culvert (1030) (Plate 43), 200 m wide and 200–240 mm high internally; the floor was composed of thin (20 mm) sandstone slabs (including at least one stone tile), the walls of roughly-coursed drystone masonry, and the roof flat slabs 50 mm thick. The interior was only partly silted up, and could be seen to continue for 1.8 m north from the edge of excavation; to the south it continued under the northern buttress, but it was not possible to see whether it was built into the buttress, or overlain by the buttress. There is little doubt that this was the continuation of the culvert beneath the northern steps of the ashpit (*Ch 5*). A slight bank of upcast on the east side contained a small assemblage of eighteenth century Delftware and later eighteenth century or later slipware (*Ch 7*).

The culvert trench was overlain by 200 mm of hard dark sand-silt with coke and clinker fragments, forming part of the trackway sequence and continuing to the east and west. This was overlain by 100 mm of stiff yellow-brown clayey sand with some brick/tile and stone (1028), which occupied the outer side of the excavation area, lensing out to the south-east under the corner of Building 1020. It merged into greyer loamy deposits to the west of the Northern Building, lensing out south beneath the north end of the later timber building (*see below*).

Beneath Building 1020, the surface of redeposited natural was exposed to the south-east, overlain to the north and west by a layer of stiff white powdery clay, with a charcoal lens at its base; it passed under trackway metalling 1032 (*see above*) to the north. Elsewhere it dipped north-west under yellow clayey sand 1028.

To the west of the corner of the Northern Building, a low mound of dark grey sand-loam with sandstone rubble and mortar lumps lay on the surface of natural-like material; its south edge dipped into and filled a shallow V-shaped gully running 1.2 m west from the building wall. The north-east edge of the mound was overlain by a thin charcoal layer, probably the continuation of the charcoal lense in the base of the white clay under Building 1020. This in turn was overlain by the more loamy continuation of clayey sand 1028.

Building 1020 (Plates 42, 43) rested on the surface of layer 1028 to the north-west, and of earlier layers to the east and south; the west wall was laid in a shallow construction trench. The walls were very similar to those of Building 945 (*Ch 6*), consisting of brick infill between postholes. A gap in the west wall was occupied by a robber trench; this overlay an irregular elongated feature including three apparent post-impressions, which may have been part of a wattle infill preceding the brick infill. However a gap at the east end of the north wall contained no robbing trench, and was probably the site of a doorway.

The postholes differed from those of Building 945 in that the post-impressions were normally clearly visible, whereas the post-pits were not (since their infills of redeposited sandy clay were very similar to the redeposited sandy clay layer through which they were cut). The clearer post-impressions were rectangular, measuring 150–160 x 90–140 mm; most were *c* 250 mm deep, but the posthole at the north-west corner was 450 mm deep. The easternmost posthole adjoined the northern buttress, and had probably formed the jamb of the doorway; a vertical recess had been chiselled into the face of the buttress just to its north.

The basal internal earth layer consisted of charcoal dust mixed with sand-silt, concentrated in the north-west corner. This was overlain by up to 80 mm of stiff brownish-yellow clayey sand with sandstone pieces, with a surface flush with those of the brick walls (which it abutted). This was cut by a slight hollow, perhaps a pathway, running north-west to south-east across the centre of the building site; remaining layers over the interior were very thin and root-disturbed, and the area had probably suffered some erosion.

To the north of Building 1020, and over surface 1028, the stratigraphy consisted of a suite of firm earth deposits, tending to merge into each other. A hard surface near the northern buttress may have been formed by trampling around the presumed doorway of Building 1020. The trackway metallings exposed in the Northeast exterior only entered the north-east corner of the area, their edges passing off site to the north-west, and interdigitating with the earth layers.

However more trackway layers were present to the north-west, along the edge of the excavation; the edges of these passed off site to the north-east, so that their intersections with the trackway sequence further east lay beyond the limit of excavation. The earliest of these metallings, over the surface of layer 1028 (which dipped quite markedly to the north-west), consisted of up to 150 mm of slag masses bedded in soft loamy sand. It was overlain by 100 mm of dark sand-silt containing coal, charcoal, tile, burnt sandstone, and slag lumps. This also lensed out to the south-east, but was bounded to the south by a sharp straight edge in line with the north wall of the Northern Building, beyond which the west edge of the excavation was occupied by a smooth surface of grey-brown sand-loam.

The straight south edge of the metalling was overlain to the west by a single large irregular sandstone slab, and was continued to the east by by a low straight scarp running to the corner of the Northern Building, and cutting through a slight mound of earlier stratigraphy to its south. To the south of the slab, a line of firebrick rubble bedded in sand-loam ran along the edge of excavation, its straight east edge aligning with that of a slot further south (*see below*). The stratigraphy of this area was shallow and root-disturbed; it passed upwards into topsoil with some rubble, and virtually lensed out completely opposite the window of the Northern Building, where topsoil lay almost directly on the surface of natural-like material.

However the stratigraphy resumed to the south, since the corner of a large bank of deposits against the west face of the furnace (*see* west centre area *below*) had spilled round the end of the northernmost buttress on the face of the furnace, protecting underlying layers from erosion. In this area, an L-shaped slot was cut into the surface of natural-like material, running west then north from the end of the buttress of the furnace. The area within this had been terraced to an almost-flat surface (whereas outside the slot the natural-like surface sloped up steadily to the south); the inner edge of the slot was up to 160 mm deep, and the upper edge up to 450 mm. The slot was infilled with rather varied loamy earth. The

slot faded out to the south, as the surrounding surfaces dipped.

Within the area enclosed by the slot, the surface was divided by a slight north-facing scarp, 0.5 m north of the internal corner of the slot, with a small circular posthole at its west end (containing slipware sherds of ambiguous dating; *Ch 7*). To the north of this lay a thin layer of coarse yellow-brown sand, whereas to the south some thin (25 mm) and fragmentary sandstone slabs may have been remnants of a floor. These layers were overlain by the north edge of the bank of stratigraphy occupying the west centre area; the base of this sequence contained large amounts of pantile rubble, which was not present further south.

The west centre area was occupied before excavation by a considerable bank of overgrown debris against the face of the furnace; in the northern 'bay' this bank was 1.5 m high, its surface lying only 200 mm below the loading hole into the furnace, whereas in the southern 'bay' it was 0.9 m high and 180 mm below the loading hole. The sequences in the two 'bays' were largely separated by the central buttress (only a few layers continuing round its end to connect them), and are described separately. The stratigraphy thinned almost to nothing along the west edge of the area.

Excavation terminated at the surface of a natural-like yellow-brown clayey sand containing charcoal and sandstone fragments, sloping gently to the north. The stratigraphy of this material was inconsistent, since in some places it abutted the face of the furnace and its buttresses, whereas in others (mainly to the south) it was cut by a narrow construction trench for the walls. It may therefore have been a mixture of redeposited and *in situ* natural.

In the northern bay, this was overlain by local deposits of angular sandstone fragments in brown sand and mortar dust, infilling the construction trench where present. These were overlain by a layer of crushed red brick mixed with sand-silt, under a layer of buff clayey sand with mortar and sandstone fragments (which passed under mortar 258 in the southern bay, *see below*). This merged east into the base of a mound, composed of alternating lenses of dark loamy sand, browner soil with sandstone, pale grey fine sand, and canary yellow sand; the mound was 300 mm high, and occupied the area beneath the loading hole of the furnace. It was overlain and surrounded by a widespread layer of pink-white mortar dust and crumbs with sandstone pieces, the outer edge of which dipped north over the beam slot in the north-west exterior (*see above*). To the south, it passed over mortar 258 in the southern bay.

Above this, there was a small mound of red silty sand with burnt stone in the centre of the area, and a separate mound of buff fine sand beneath the loading hole. These were surrounded by 200 mm of dark sand-loam, under a consistent layer of white mortar crumbs mixed with sand-silt and small sandstone pieces, up to 150 mm thick. It could be tentatively correlated with a similar layer in the southern bay, whereas the overlying stratigraphy was very different to that further south.

This upper stratigraphy started with a 100 mm layer of soft dark sand-silt with mortar flecks, overlain by up to 250 mm of loose pale red sand with burnt sandstone and firebrick rubble. This in turn was overlain by a steep bank of firebrick and burnt cementation chest sandstone in a matrix of soft dark earth, 500 mm thick against the face of the furnace (Plate 45). The bank had a very straight west edge in line with the ends of the buttresses, which may have been revetted (it was severely root-disturbed, lying immediately beneath topsoil).

The stratigraphy of the southern bay was shallower and simpler. The earliest feature over the natural-like base was ridge of charcoal fragments, in a matrix of charcoal dust mixed with sand-silt, running north-west from the southern buttress and up to 0.3 m high. The inner edge of this bounded a continuous surface of hard solid brown-white mortar with small sandstone lumps (258, *see also above*), which had set *in situ*. This occupied the whole bay, and adhered to the face of the furnace, forming a local rendering. Its flat surface contained three cast-in slots (Plate 44), all with vertical sides and flat bases; two were rectangular (measuring 1.08 x 0.10 m by 0.02 m deep, and 0.96 x 0.22 m by 0.05 m deep), and the third curved (measuring 1.35 x 0.35 m by 0.04 m deep). These had no apparent function, and may have been merely the sites of planks laid on the pre-existing surface, fortuitously preserved by what may have been an accidental mortar spillage rather than a deliberate floor.

The mortar surface was overlain by a stratigraphy dominated by heaps of sand (successively bright yellow and coarse, light brown and fine, brown and fine, and red-brown and fine) beneath the loading opening, interdigitating with layers of mortar dust and lumps with variable earth contents, passing up into less mortary sand-loams and sand-silts with some rubble content; this sequence extended to the base of topsoil.

The topsoils of the area consisted of dark crumbly loam with firebrick, pantile and stone rubble close to the furnace, passing into black silt-loam along the west side (the colour probably deriving from a high charcoal dust content); both topsoils were intensely affected by root activity from scrub growth.

The south-west exterior area was very simple stratigraphically. Before excavation, the northern part of the area was fairly flat; the pathway trench beside the Southern Building (*see* southern exterior description) extended south at a similar level, while to its west the ground sloped steadily up to the south. The stratigraphy of the pathway trench has already been described (*Ch 6*), and the area to its west was virtually featureless; remaining features were therefore concentrated at the north end of the area.

Over most of the area, excavation halted at the surface of yellow-brown clayey sand. This material was investigated in a test trench, 0.4 m deep, running west from the junction of the Southern Building with the southern buttress of the furnace. The west end of this trench exposed an irregular surface of hard brown clayey sand with cobbles; over and east of this lay a layer of stiff yellow clayey sand with a few cobbles. Both are interpreted as *in situ* Boulder Clay. They were overlain by an average of 150 mm of hard yellow-brown sand with scattered pebbles. The construction trench for the furnace was cut from the base of this layer, but the trench for the Southern Building was cut from within the layer, which is therefore interpreted as a mixture of *in situ* and redeposited or trampled weathered natural.

The southern buttress of the furnace was trench-built, as were the footings of the Southern Building wall; above footing level the wall had been built free-standing, the construction trench being infilled with soft brown clay.

In the north-east corner of the area, a small rounded pit was cut directly into the weathered/redeposited natural; it measured 0.7 x 0.6 m by 0.22 m deep, and was infilled with yellow-brown silty sand over charcoal debris. The pit and surroundings were sealed by a more extensive layer of charcoal-rich loam, up to 150 mm deep; the relationship of this deposit to the edge of the west centre stratigraphy was unclear due to severe root disturbance. Part of this was overlain by an area of laid firebrick and stone rubble in the angle between the Southern Building and the furnace; this had probably been laid to consolidate the surface, for reasons which are not apparent.

To the south of these layers, an infilled trench ran north-west from beneath the sill of the blocked doorway of the Southern Building. It contained both a rough brick drain and an iron pipe, both falling to the north-west, and fed from the iron trough within the agricultural conversion of the Southern Building (*Ch 6*). Elsewhere the surface of yellow clayey sand was directly overlain by topsoil, consisting of 150–400 mm of dark loam with some rubble.

The outlying trenches

The remaining areas were excavated, or recorded during work by English Heritage's Directly Employed Labour, as unavoidable interventions necessary for the opening of the monument to the public, rather than for the understanding of the monument. They are described very briefly, from north to south.

Two pits, each *c* 1 m square and 1.8 m deep, were dug for rubbish disposal to the north-west of the furnace complex. They were located within the edge of deciduous woodland, on a slight ledge at the top of the seemingly-natural slope from the furnace area down to the Derwent floodplain. These pits terminated on the surface of pale brown silty sand with small gravel, occasional large rounded cobbles, and rare coal flecks, assumed to be natural Boulder Clay. This was overlain by 100–200 mm of very hard gravelly sand concreted with ironpan, containing pieces of burnt and glazed sandstone and slag. This was succeeded by 200 mm of soft grey sand-silt with ironpan streaks, under 200 mm of grey and brown hard silty sand with silt and sand lenses and large lumps of rusty slag. This was abruptly overlain by 0–150 mm of firm silty sand with gravel and lumps of friable slag, under 200–300 mm of mottled silty sand with sand and gravel lenses. This in turn was overlain by 800 mm of dark sand-silt, grading up into dark grey loam topsoil. The stratigraphy indicated an unexpected amount of anthropogenic deposition, during and after the use of the furnace; the majority of this sequence may have dated from after closure of the furnace, and have been produced by hillwash from the (cultivated) enclosure around the furnace.

Two sets of trenches were excavated (mechanically, with archaeological observation) along the west side of the site. One of these was for drainage, running north-west then north from the Southern Building to discharge on the steep wooded slope north-west of the furnace; the other was for mains services, running south-east then south from the Southern Building (with a spur to the custodian's hut site, south south-east of the Southern Building), and extending along Forge Lane. Trenches were typically 0.5 m wide and 0.7 m deep.

Most of the northern trench ran close to the boundary hedge between the furnace enclosure and Forge Lane. Its base was cut into soft grey-yellow sandy clay, unlike the normal Boulder Clay natural, and perhaps a hillwash deposit filling a hollow in the natural topography. This was overlain, in the northern part of the trench, by a sequence of trackway metallings (consisting successively of concreted slag and metallurgical debris, very hard rust-stained sand, and very hard brown sand), interleaved with softer earth and sand deposits. The surfaces dipped and thickened to the north; the south edge had progressively migrated north during use, from an initial position opposite the north-west corner of the Northern Building, and the track had run more-or-less at right angles to Forge Lane. The south end of the trench crossed diagonally from the hedge to the blocked western doorway of the Southern Building; in this length soil horizons directly overlaid natural clayey sand, the only features being the continuations of the agricultural drain and pipe running from the doorway (*Ch* 6).

The north end of the southern trench ran within the previously-excavated area, and revealed no new information. It then ran from the south-west corner of the excavation to the edge of Forge Lane at the south-west corner of the furnace enclosure. From 7 m to 20 m from the corner of the excavation, it cut across a wide gentle hollow in the natural topography (deeper than the trench). Where exposed, the bottom of this hollow contained thin stratigraphies of mortar (to the north) and hard grey sand containing forge slag, fritted sand, and vitrified material (to the south); above these, the whole hollow was infilled with dirty brown sandy clay. To the south of this, natural yellow-brown clayey sand was overlain by thin deposits of earth and metallurgical debris, at least one of which formed the metalling of a path (visible as a slight hollow in the modern ground surface).

The remainder of the trench ran along the east side of Forge Lane. For the first 14 m, the Lane is narrow and sloping, and probably on a pre-nineteenth century line; further south it is straight and flat, and (from historical evidence) was laid out in or after the 1830s. Beside the sloping part, the trench exposed a deep stratigraphy of earth layers with forge slag and other debris, including one strip of large rubble and slag blocks; the inclination of these deposits suggested that they were within a roadside ditch. Further south the only stratigraphy was a layer of iron-rich debris with forge slag blocks, directly under topsoil; even this died out southwards.

The remaining area of excavation lay to the south of the Southern Building, where an area 6.8 x 4.8 m was excavated to prepare the site for a custodian's hut. The site lay just to the north of a stone building foundation, and to the west of a scarp into the rising ground surface; it looked like a northward extension of the building platform for the stone building.

The excavation was taken down c 250 mm, to a hard surface, with limited deeper trenching for the hut foundations and surfaces.

At the north-west corner, the surface of natural yellow sandy clay was directly overlain by a surface of hard red sand, becoming earthy at its edges; the surface of natural was reddened beneath this, presumably by *in situ* burning. Within the central part of the area, the deeper trenches revealed that the yellow clay surface included redeposited material in backfilled features; these could be traced on the surface, and could not be interpreted. One definable pit was present, protruding from the south edge of excavation; it was a pit filled with pantile rubble in soft dark sand-loam, perhaps a soakaway.

Over most of the area, natural was directly overlain by a surface of small rounded sandstone rubble, with varying amounts of pantile and mortar, bedded in hard grey earth. An irregular structure of sandstone bedded in hard cream-white mortar, measuring 0.67 x 0.64 m, protruded from the surface near the north-east corner; it was probably a pier base or a post-pad. A second similar feature was partly exposed on the south edge of excavation (3.6 m south of the first, centre-to-centre).

The rubble was overlain by topsoil, with some local rubble layers and small features (including a deposit of Type A slag in the south-west corner). These layers contained abundant early twentieth century pottery (*Ch 7*).

This area is interpreted as the site of an open-sided building with a rough rubble floor, beside the stone building to the site. The pottery over its floor at first sight implies a twentieth century date, but strictly provides only a *terminus ante quem*. Since the building is not shown on any OS map edition, it is likely on balance that the building was used and abandoned before the 1850s, the pottery deriving from much later rubbish tipping. The dirty clay infill of the hollow to its west (revealed by the service trench) may well have been upcast from the digging of the platform for the building.

7

THE FINDS

The excavation produced a very large assemblage of finds, dominated in terms of bulk by firebrick and process residues. Unfortunately only a very small proportion of this assemblage was stratified within the industrial use of the furnace, the vast majority being recovered from contexts post-dating closure of the furnace (though naturally including material of earlier origin). This resulted in part from excavation strategy (in that post-abandonment deposits were systematically removed, whereas only limited excavation of use-period stratigraphy was undertaken, and use-period waste and rubbish tips lay outside the area of excavation), and partly from depositional factors; it would appear that the furnace and its immediate environs were kept clean during use, with only very limited deposition of rubbish and process waste, whereas after closure the site was used for rubbish tipping, while the collapse rubble that accumulated contained large amounts of building materials (hardly surprisingly). The bulk of the artefact assemblage therefore represents post-abandonment rubbish unrelated to the use of the furnace; building materials and process residues did derive from the use of the furnace, but were largely recovered from poorly-stratified contexts.

Only very brief reports on most categories of finds are therefore presented in this report, the bulk of the information being retained in archive catalogue form. The assemblage can be broadly divided into artefacts, building materials, and process residues, but these divisions are not always clearcut; it is therefore categorised by material rather than strictly by function. The bulk of the finds have been catalogued and identified by Sandra Hooper, the assemblage from the 1991 excavation being added by Jenny Vaughan. Specialist reports have been prepared by Edgar Bolton (clay pipes), Margaret Chard (glass), Louisa Gidney (bone), and Gerry McDonnell (slags and metallurgy). Of these, Mc Donnell's and part of Chard's work are published in full below, the remaining contributions being retained in the Site Archive. Staff of the Victoria and Albert Museum (V & A) advised on non-local parallels and dating for the pottery. Jenny Vaughan has advised on various aspects of interpretation and presentation.

Artefacts

The pottery

The vast majority of the pottery recovered was of later nineteenth or twentieth century date, the assemblage adding little to those already published from the region (*eg* for local red earthenware see Bown and Nolan 1990, 107–111). Quantities of porcelain were very small. The stoneware assemblage was also small, but included single sherds of patterned Staffordshire salt-glaze plate, and of Staffordshire scratch-blue 'tavern ware', both of mid-late eighteenth century date (Jennings 1981, 223, 227). At the other extreme, a jam-pot from the late drain in the Southern Building (*Ch* 6, 7) is dated by the V & A to *c* 1900. The tinglaze assemblage included seven sherds of eighteenth century Delftware, some with eighteenth century parallels (Dawson 1971).

The white earthenware assemblage was larger, the majority having blue-on-white decoration; dateable material was of mid to late nineteenth century date. Probable sources include Sewell & Donkin's St. Anthony's Pottery, John Carr's Low Lights Pottery (North Shields), and C T Maling's Ouseburn and Ford Potteries (Bell 1971).

The redware assemblage was predominantly of Sunderland Ware and similar wares (redwares with brown glaze externally, and yellow glaze internally). The small slipware assemblage presents problems of dating; nationally most of the wares would be accepted as of mid to late eighteenth century date (trailed slipwares *cf* Ashurst 1987, 209; combed slipwares *cf* Jennings 1981, 105), but Scott's Pottery, Sunderland, produced similar wares until the late nineteenth century (Baker 1984, 50). The stratigraphy and associations of some (but not all) of the

Derwentcote material support an eighteenth century dating. A 'brown earthenware manufactory' was located at Derwentcote in 1757 (Buckley 1927, 76–7); since the products of this kiln have not been characterised, it is impossible to say whether they are represented in the excavated assemblage.

The assemblage from the excavation for the custodian's hut (*Ch 6*) contained large amounts of late nineteenth-earlier twentieth century pottery, including a 'Bungalow' ceramic footwarmer and a green-glazed 'woven basket' style jar, both closely paralleled in the Harrods catalogue for 1929.

The clay pipes

The excavation yielded a small assemblage of clay tobacco pipe fragments (31 fragments, from a maximum of 29 individual pipes); one fragment in reduced fabric, the remainder in generic white pipeclay. There were no complete bowls or determinable bowl forms. The few dateable pieces were of late nineteenth–early twentieth century date, and came from post-abandonment contexts. These included seven fragmentary maker's stamps, from at least five different dies, notably those of local manufacturers F J Finn and J Hardy (providing respective *termini post quem* of 1891 and 1887 (Jarrett 1964, 259); Parsons 1964, 251). One of the former derived from a fill of the ashpit. Remaining fragments were undiagnostic; on stratigraphic grounds some are likely to be of earlier date, but this could not be confirmed by positive identification.

A detailed report is filed in the site archive (Bolton 1990).

The glass (Figs 13, 14)

Margaret Chard and David Cranstone

Glass was recovered from contexts of all major phases of the site, providing useful dating evidence. The largest assemblages were associated with the abandonment of the furnace. Five major categories of product were present: bottles, window glass, roof glass (glass pantiles), pharmaceutical wares, and clear glasswares (tablewares etc). The catalogue is based on these categories, subdivided by 'metal' and by form/function (Figs 13, 14).

The 'metals' include a basic green metal (little changed since the mid seventeenth century), an almost-clear metal with a bluish-green tint (developed in the earlier nineteenth century), and a clear lead glass (developed at the end of the seventeenth century, but a rare luxury until the later nineteenth century). Forms are related both to developments in manufacturing technology (*eg* moulding and mechanised production) and changes in sale and manufacture of products (*eg* pasteurisation of beer,

and sale of medicines in multiple doses). The type series of bottles (B1–10; B7, B9 and B10 not illustrated) and of pharmaceutical wares (Ph1–5; Ph4, Ph5 not illustrated) are shown in Figs 13, 14, excluding very late types, and those of which only fragments were recovered.

Quantities of eighteenth century material were small. The only closely dateable vessel was a wine bottle (type B1), the form of the neck suggesting a date of 1730–45 (Ross 1982). Fragments came from several well-stratified contexts in the southern exterior (*see Ch 6*). Other green glasswares (from this sequence and elsewhere) may have been of similar date, but were not diagnostic.

Three small fragments of flat glass, 1 mm thick, in green metal (W1), were retrieved from stratigraphically early contexts. These can be identified as locally made 'broad glass', used at the cheaper end of the market from the seventeenth to the early nineteenth century (Ross 1982). The bulk of the window glass however was of flat polished glass in clear blue metal with a bluish-green tint (W2), 1.5–2.5 mm thick. The colour suggests use of a soda silicate flux (Butterworth 1948); this improved crown glass was patented by Attwood of the Tyne Glass Company in 1817, and improved by Isaac Cookson in 1831. It went out of production at the end of the nineteenth century. A deposit of this glass was recovered from beneath the window in the west wall of the Southern Building.

A large proportion of the total glass assemblage came from the infills of the ashpit. This included large amounts of type W2 in various thicknesses (including two fragments with floral stencil frosting), and of mulberry coloured cast plate glass with moulded daisy pattern (W4). The former may have derived from windows on the site, but the latter clearly originated from a domestic building of some pretensions (possibly Derwentcote House). The assemblage also included large quantities of bottle glass, of two types (B3 and B5) both in the same metal as W2. The B3 bottles had collared lips to take a tied-on cork, implying use for unpasteurised beer. Pasteurisation was introduced in the 1870s; tied corks were rapidly superceded by internal screw stoppers (patented 1872) and crown corks (patented 1892) (Meigh 1972). The B5 assemblage consisted of gin bottles marked 'W & A GILBEY'; this company was founded in 1857, had agents in the North East from at least 1869, and became 'Limited' on incorporation in 1893. The bottles had clearly been broken on site, and the combination of types suggests a date between the late 1860s and the late 1870s. Since some fragments had been exposed to heat after breaking, it is tempting

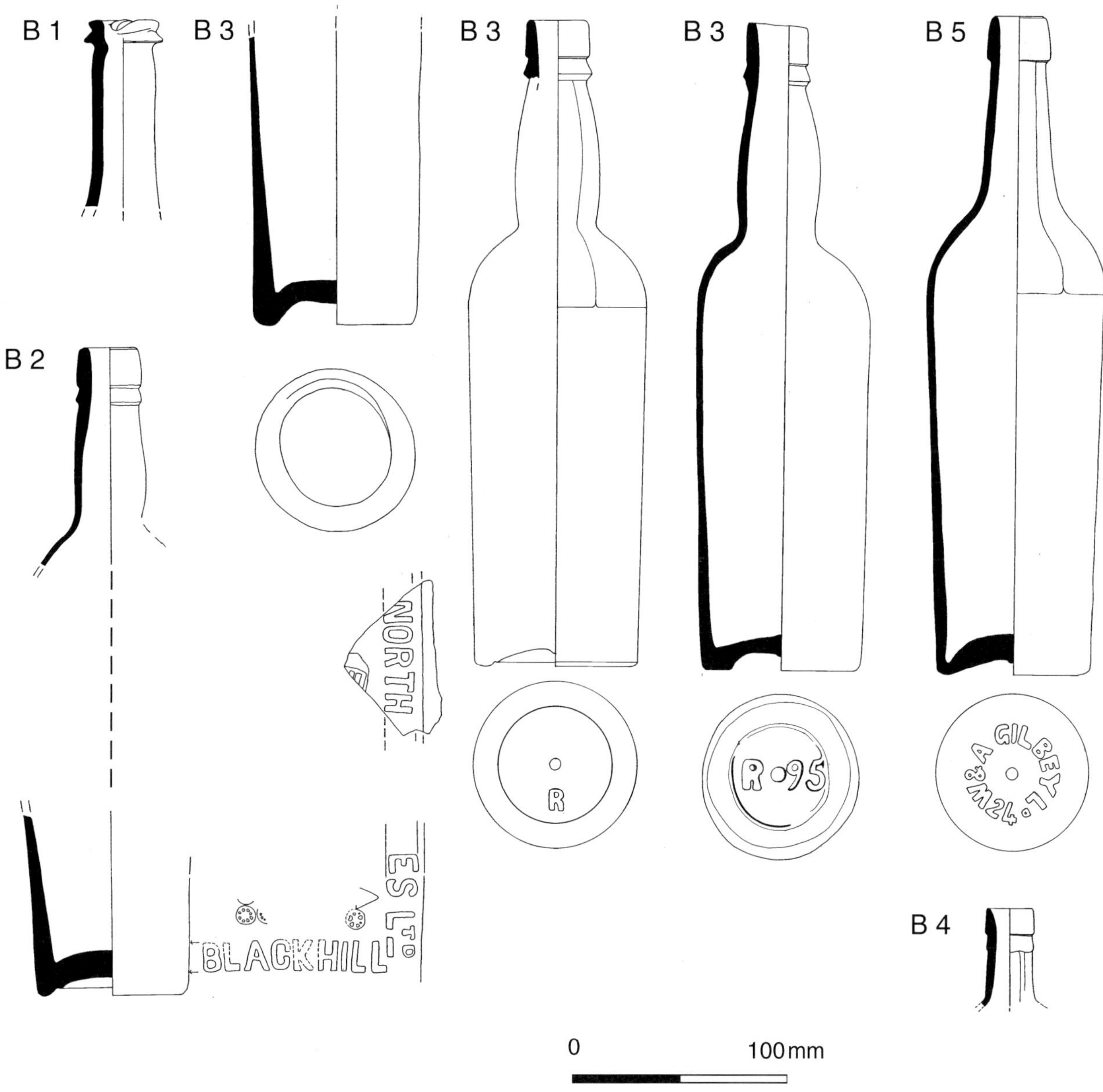

Fig 13   Glass: bottles types B1 – B5

to see this as the remains of a closing-down party, when the ashes were still hot.

The glass pantile fragments were recovered from post-closure and modern contexts, indicating the period at which the roof deteriorated, rather than when they were introduced to the site; four to six tiles were represented. Glass tiles were patented by J Bowron (a Tyneside manufacturer) in 1855 (Ross 1982); they may therefore have been used in the final industrial phase of the site, or in the agricultural conversion.

The pharmaceutical wares included three single-dose phials (Ph 1 of seventeenth-eighteenth century date from an early context in the Southern Building, and Ph 2 and Ph 3, both probably of nineteenth century date and from post-closure contexts), and a range of multiple-dose bottles (probably all of nineteenth century date) from late contexts.

A detailed report is filed in the Site Archive (Chard 1990).

The ironwork (Figs 15 – 24)

The ironwork assemblage was relatively large and varied. A high proportion consisted of structural fittings, and some specimens can also be considered as process residues/raw materials. Classification is

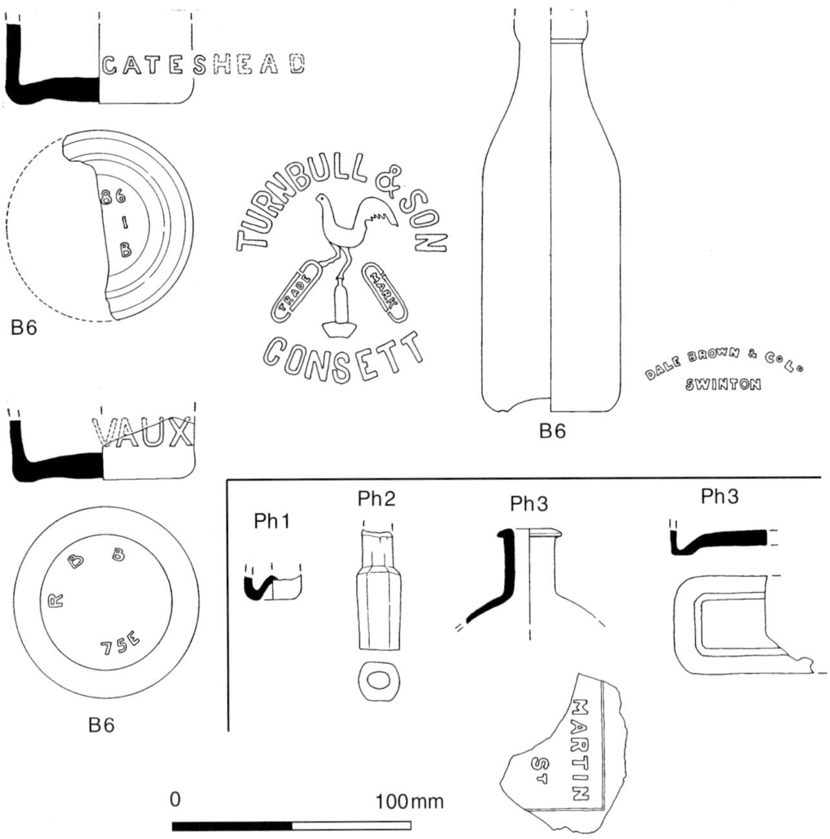

Fig 14   Glass: bottle types B6, B8; pharmaceutical vessel types Ph 1–3

in some cases dependent on interpretation, but the following broad categories can be identified. A high proportion of the total assemblage consisted of unidentifiable fragments.

Structural fittings

a)   Door fittings (Fig 15) included three 'claspers' similar to that found *in situ* holding a scarfed-in repair to the jamb of a door frame (*Ch 5*) (no 26), hinges of four types (nos 26, 28, and 29; a plain rectangular plate-hinge is not illustrated), two keys (54, 55), and three types of latch and latch-lifter (nos 24, 25, 30). One hinge came from the ash in the ashpit (implying use of the frame as kindling wood), the remainder being from post-closure contexts.

b)   Small fittings, including several wall hooks and brackets (not illustrated). One length of chain attached to an iron spike was presumably wall-mounted; two other lengths of chain could have served various purposes.

c)   Rainwater goods (Fig 16) included guttering and a supporting bracket (56 and 56a), from contexts immediately outside the furnace buildings. Fragments of piping of various diameters may also have been from gutters or downpipes.

Individual large objects (Figs 17, 18) included a grille (60) from an infill of the southern ashpit; part of a firebar (74), from a raft of rubble in the soil layers east of the furnace; a trapezoidal plate (62), recovered from an infill of the northern ashpit and almost certainly a part of the floor over this; a heavy #-shaped iron frame (61) from superficial rubble over the northern ashpit, almost certainly the frame for the firedoor of the firegrate; and a thin plate with two attached hinges and a cast-in handle (63), from the same context and probably a heat-proof firedoor from the firegrate or from the main opening into the cone (*Ch 5*), just above the findspot.

Assorted nuts, bolts, and washers, and some of the rods and bars (as categorised below), may have originated from structural fittings, but their function cannot be identified (2).

Slabs

Eight triangular or rectangular slabs of iron, varying in thickness from 0 mm to 25 mm, were recovered from various contexts and areas. These may well have originated as flooring, in the Southern Building

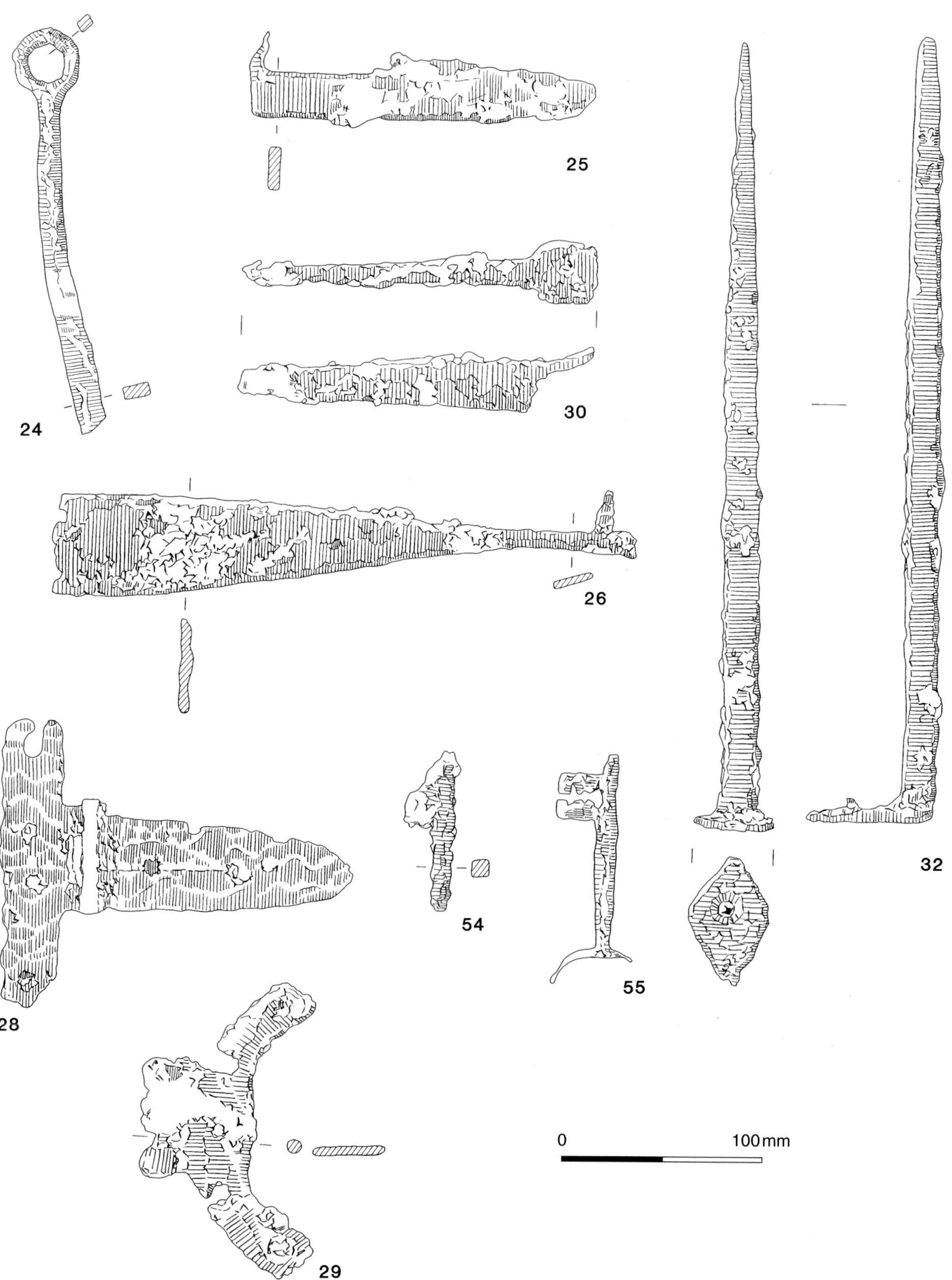

Fig 15   Ironwork: door fittings

79

or elsewhere. In addition, two large semi-circular cast iron plates were found re-used in a post-closure floor of the Southern Building (*Ch 6*). One of these was removed and drawn (Fig 19), before being reinstated in the conserved floor. The annular wear pattern on these plates suggests that they had formed the base of an edge-runner grinding mill, perhaps for charcoal.

Nails

Approximately 178 nails were recovered, of which some 25% were fragments, and others were severely corroded. The better-preserved examples can be classified into the following types (Figs 20, 21):

> N1. Truncated pyramidal head, square shaft, pointed. 8 examples.
> N2. Domed head, one face continuous with square or rectangular shaft, not pointed ('cut nails'). 21 examples.
> N3. Pyramidal head, rectangular shaft, pointed. 2 examples.
> N4. Rectangular flat head, projecting to one side from rectangular shaft, not pointed ('cut floor nail'). 1 example.
> N5. Circular flat head, with circular, square, or rectangular shafts, pointed. 6 examples.
> N6. Circular domed head, pointed;
> a) Circular shaft. 2 examples.
> b) Rectangular shaft. 2 examples.
> c) Square shaft. 7 examples.
> N7. Triangular shaft. 2 examples; one with circular domed head, pointed, the other seemingly pointed at both ends.
> N8. Slightly domed square head, rectangular shaft (no points surviving). 2 examples.
> N9. Miscellaneous (damaged). 18 square shaft; 11 circular shaft; 25 rectangular shaft.

There was considerable variation in size (and no doubt in function) within each type; due to the high proportion of incomplete examples no systematic analysis has been attempted.

Tools (Figs 22, 23)

This category included a hammer head (33), three files (34, 35, 36), a trowel (37), a scraper (38), an adze (39), and three chisels (one large, perhaps a 'jumper' for manual rock-boring (43)). The adze came from the trampled base of the pathway trench (*Ch 6, 14*); the remaining items were from post-closure or poorly-stratified contexts.

Three types of larger object may also be considered here. The first consisted of a rectangular bar with plate-like terminal (41); three examples were found. The second consisted of two rods, sub-circular in section, with evenly curved ends very similar to each other (44, 46). The third consisted of three rods with strongly-hooked ends (47). All were from post-closure contexts, and may have been industrial or agricultural tools, or structural fittings.

Remaining rod-like objects cannot be reliably categorised. These were circular or square in section, measuring 8–30 mm, and of varying (normally incomplete) length. Some were tapered or pointed, suggesting interpretation as spikes; the remainder probably included broken parts of both tools and structural fittings.

Similar problems arise with the various fragments of iron strip and sheet. One strip can be identified as a tyre or hoop, from a wheel or barrel of 25 mm diameter, and pieces of curved sheet may have originated from a bucket.

Bars are considered separately below, but a group of seven D-shaped objects, rectangular or (in one case) trapezoid in section, may be mentioned here. These varied in size (from 150 x 150 x 100 mm to 60 x 50 x 20 mm) and precise shape; some or all may have been flat irons.

Bars

A large number of bar-like objects was recovered; some of these are likely to been parts of tools or structural fittings, whereas others (notably the plain rectangular bars) may well have been the wrought iron raw material for the furnace, or the blister steel product. One complete bar, and five broken lengths, were semi-circular or segmental in section, and two were trapezoid; these latter may have been firebars, though their sections differed from the identifiable example.

A total of 8 bars and parts had rounded ends; one may have been a firebar, the remainder being relatively elongated rectangles in section. Another group of six specimens had bevelled ends, and may have been tools.

The remainder of the assemblage consisted of plain rectangular bars; 41 measurable, and 70 unmeasurable, fragments were present. The cross-sections show no evidence of precise standardisation, though most were relatively wide and thin. Widths varied from 20 mm to 105 mm, and thicknesses from 10 mm to 25 mm, section of *c* 80 x 20 m being common. This accords with dimensions of 2–3" (50–90 mm) by $^3/_8$–$^3/_4$" (10–60 mm) quoted as the normal range by Barraclough (1984a, 36) for Swedish bar iron for steelmaking. One bar was stamped with an L in a circle (Fig 25), the 'Hoop-L' mark widely preferred for steelmaking (Barraclough 1984a, various refs).

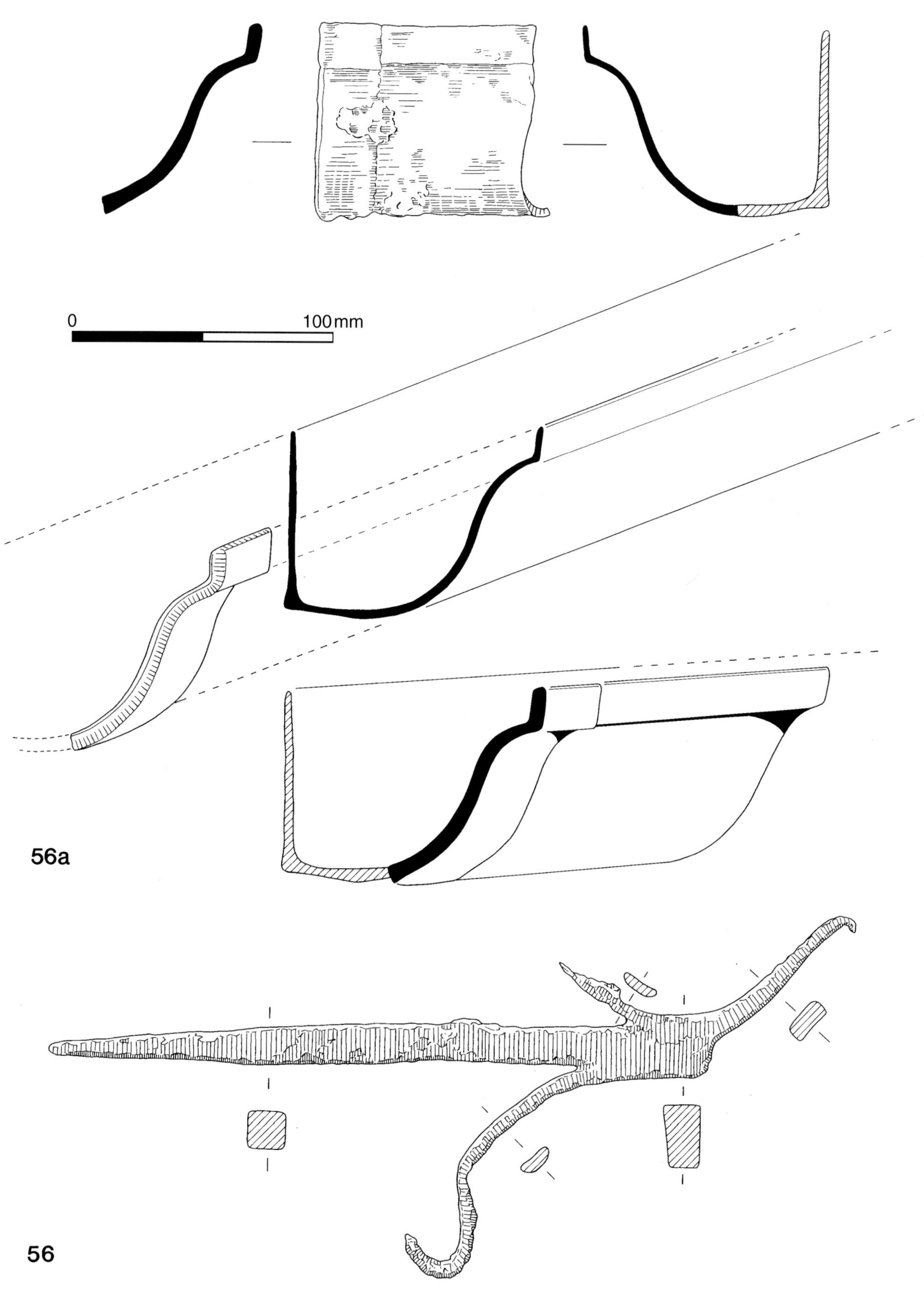

56a

56

Fig 16   Ironwork: rainwater goods

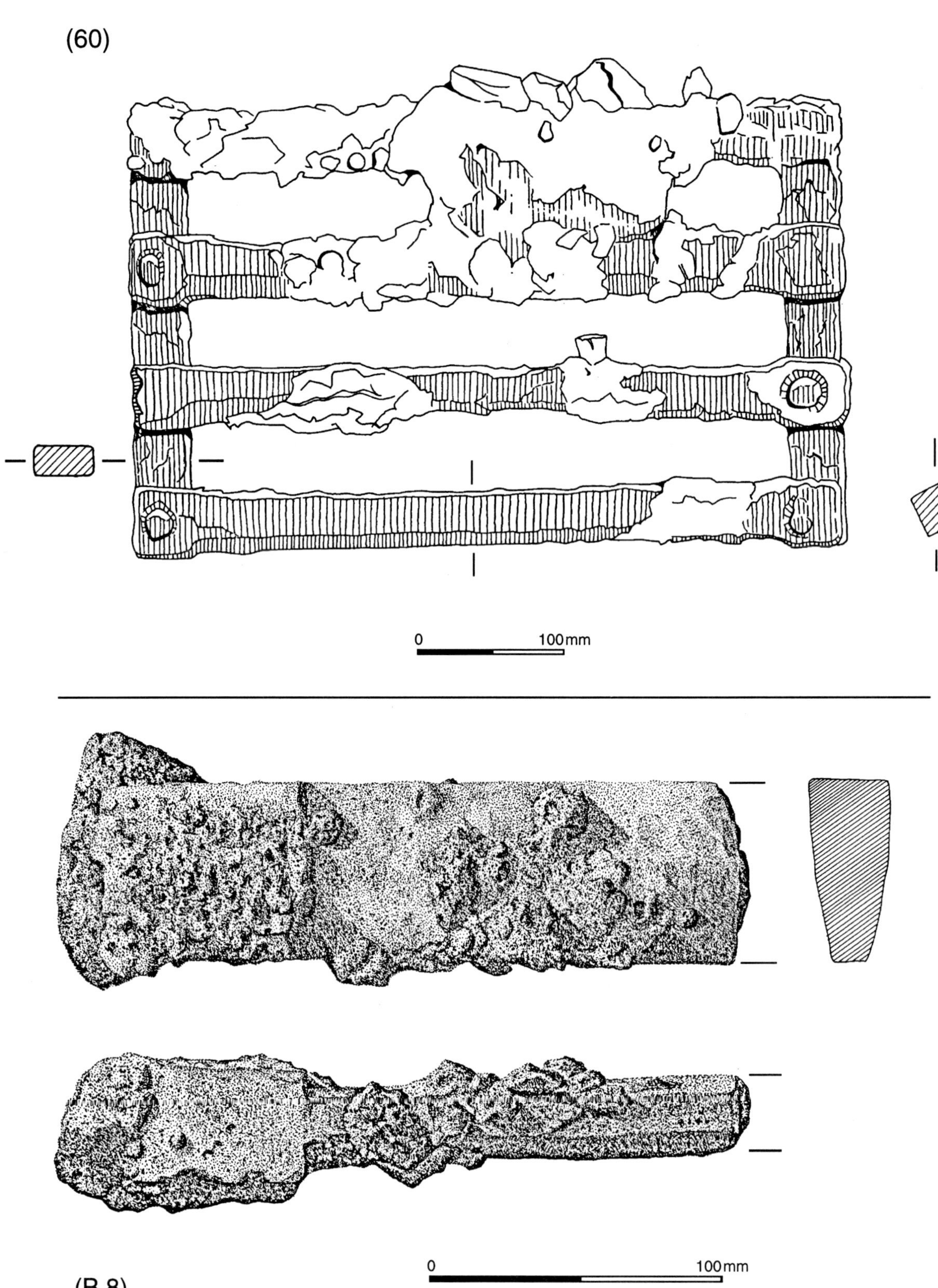

Fig 17   Ironwork: iron grid (60), firebar (B8)

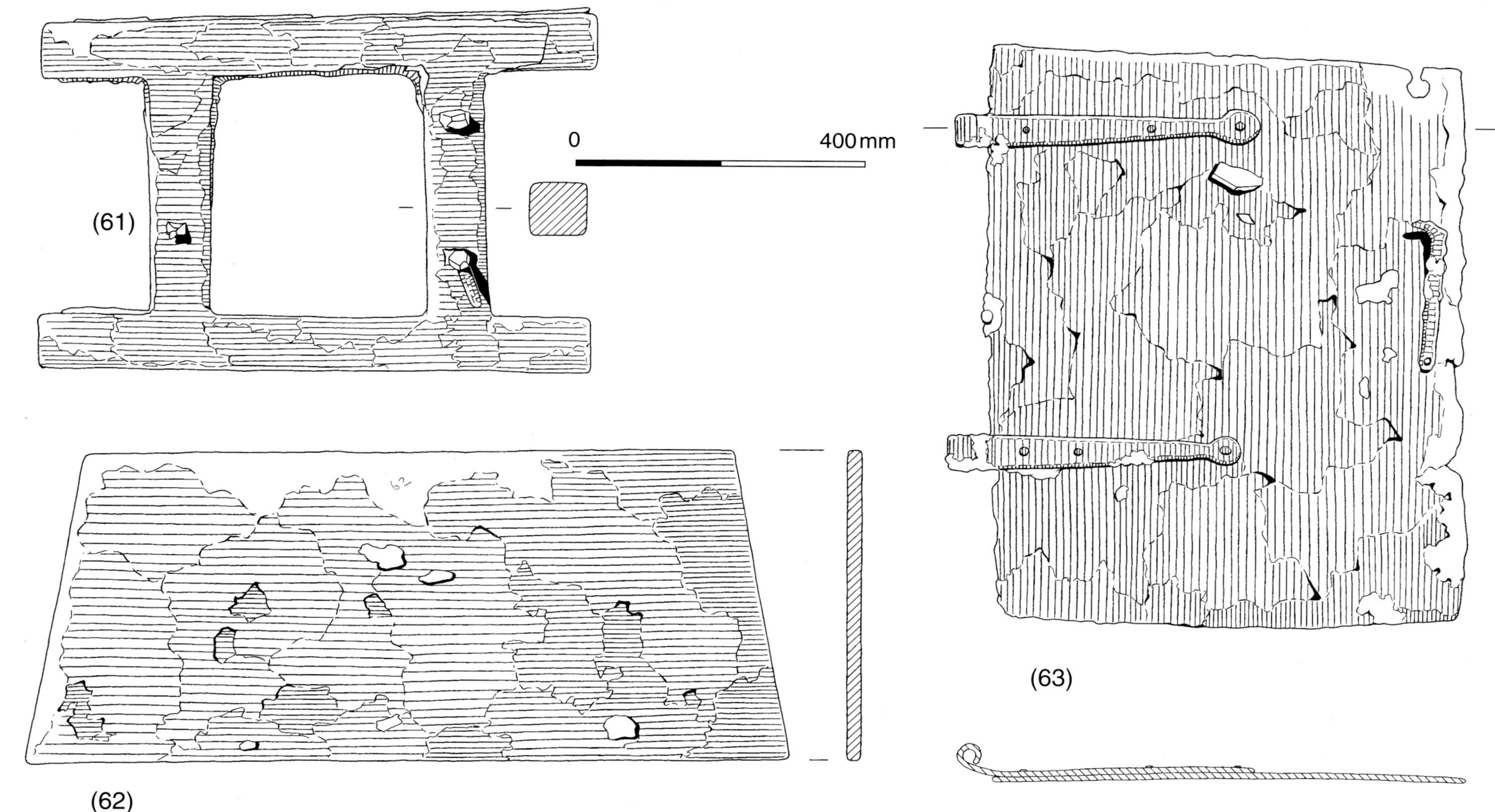

Fig 18   Ironwork: frame from entrance to firegrate (61); cover plate from ashpit (62); iron firedoor (63)

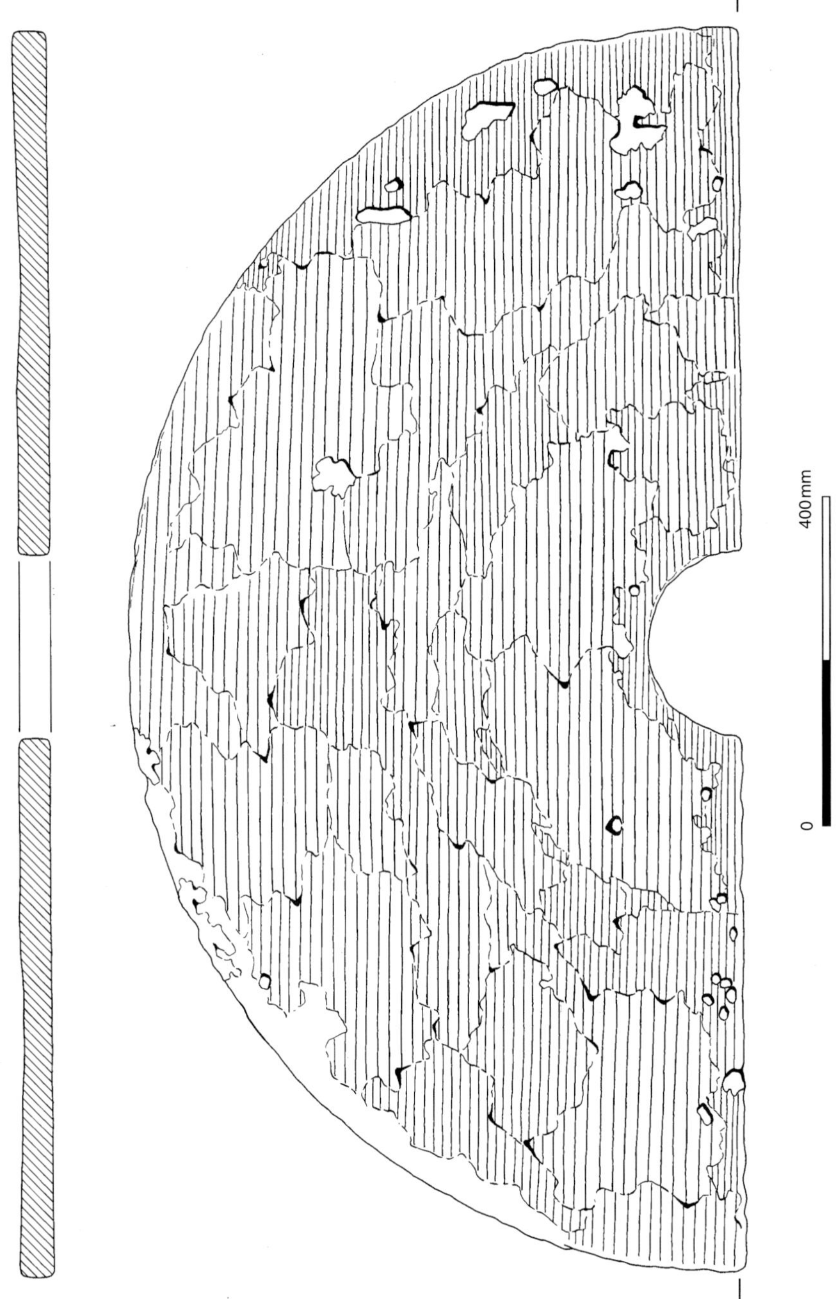

Fig 19 Ironwork: semi-circular plate

Thermally-affected material

A consistent component of the assemblage was fragments of iron or steel, almost invariably 5 mm thick, with a distinctive silver-blue to silver grey metallic sheen; *c* 130 fragments were recovered, the majority from contexts in or near the ashpit.

Other metals

Lead
One piece of perforated and folded lead sheet and one fragment of lead strip were recovered from late contexts in the south exterior.

Copper alloy
Four buttons, four discs (two with iron corrosion products adhering, two nails, and four other objects (two of them modern-looking) were recovered; the buttons were stratified in the sequence of the east centre area (*Ch 6*), and the nails in trackway surface 973 of the north-east exterior (*Ch 6*).

Coins
The only coins recovered were a penny of 1860 from the Eastern Building (*Ch 6*), and a halfpenny of 1861 from topsoil.

Miscellaneous artefacts

Other finds included one bone handle, one mother-of-pearl button, fourteen pieces of artefactual wood (none of wider interest), one leather shoe sole from a post-abandonment context, and three pieces of leather sole from an industrial-period context.

Building materials

Pantiles

Large quantities of pantile were present in most of the post-closure and recent contexts, deriving from collapse of the final roofs of the buildings. This material was normally discarded on site, pantile only being retained where the fabric was unusual, and from contexts pre-dating the closure of the furnace (in order to assess the evidence for earlier roofing materials).

The pantiles were normally in a fine red fabric, sometimes with visible (but fine) sand tempering. Tiles were consistently $^5/_8$" (16 mm) thick; the outer face was smooth, whereas the inner was rough and sandy. A few examples (all from post-closure contexts) were fired to a grey colour.

Although the overwhelming bulk of pantile was from post-closure contexts, small quantities were present in various contexts from the later part of the industrial sequence, and also from pit 1138 in the Southern Building (*Ch 6*), and from the fossil soil sealed beneath the stratigraphy at the east end of the Southern exterior (*Ch 6*). This indicates that pantile was present at the site from the beginning, and, in conjunction with the pantile packing in the pointing of the northern buttress, and the rarity of other roofing materials, suggests that the roofs had been pantiled from the outset.

Other roofing materials

The presence of glass pantiles has already been noted (*see above*). One definite, and one possible, fragment of sandstone roof tile were recovered, both from post-closure contexts. In addition, early culvert 1030 at the north end (*Ch 6*) was floored with sandstone tiles (which were not removed). True roofing slate (probably of Lake District or Welsh origin; a geological identification has not been obtained) occurred as a lining in a late slot in the Southern Building, which may or may not have pre-dated closure of the furnace (*Ch 6*), and (in very small quantities) in the ashpit and in external topsoil. These quantities do not suggest any use for roofing within the furnace complex, and may have been derived (along with other debris, noted above) from Derwentcote House.

Housebrick

Appreciable quantities of housebrick were recovered, again largely from post-closure contexts; the quantity is slightly surprising in view of the very limited use of housebrick in surviving structures and features. The majority was in relatively soft sandy handmade fabrics, fired red; overfired or vitrified specimens were common. A small assemblage was recovered from early contexts in the Southern exterior; these were in a sandy orange fabric with sand and black grit inclusions, with darker red to maroon surfaces, vitrified black in some cases. Dimensions varied on the few measurable specimens. Small amounts of brick in a hard red sandy fabric were also recovered from early contexts in the southern Building. Small amounts were also stratified in the north-east and north-west exteriors; these may have originated from Buildings 945 and 1020.

Firebrick

Large quantities of firebrick were recovered, overwhelmingly from post-closure contexts; this is hardly surprising in view of the large amounts used in the internal linings of the furnace. From post-closure contexts, only specimens showing two original measurements, or any part of a stamp, were

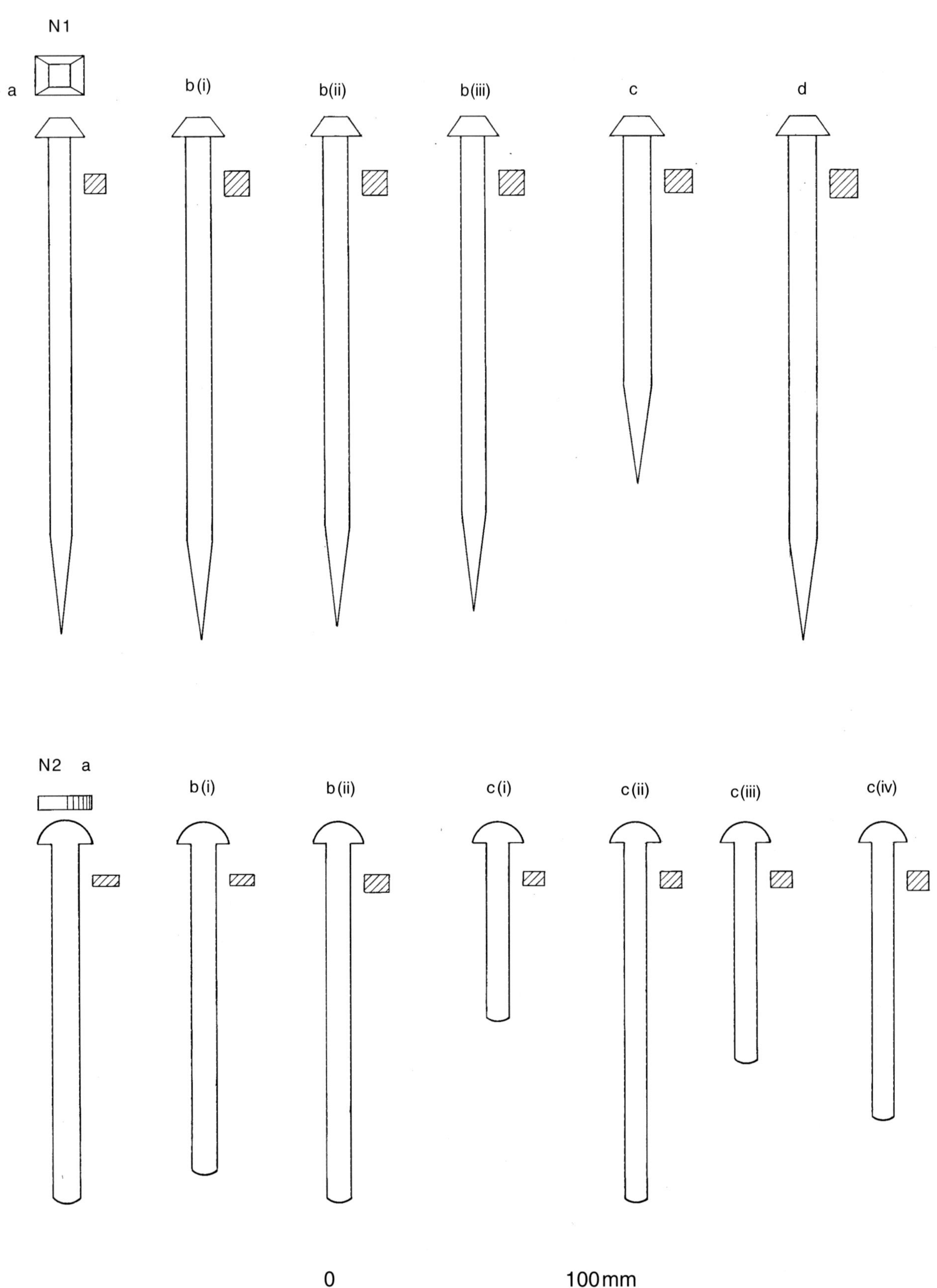

Fig 20 Ironwork: nail types N1–N5;
Fig 20a (opposite page) nail types N1–N5 continued

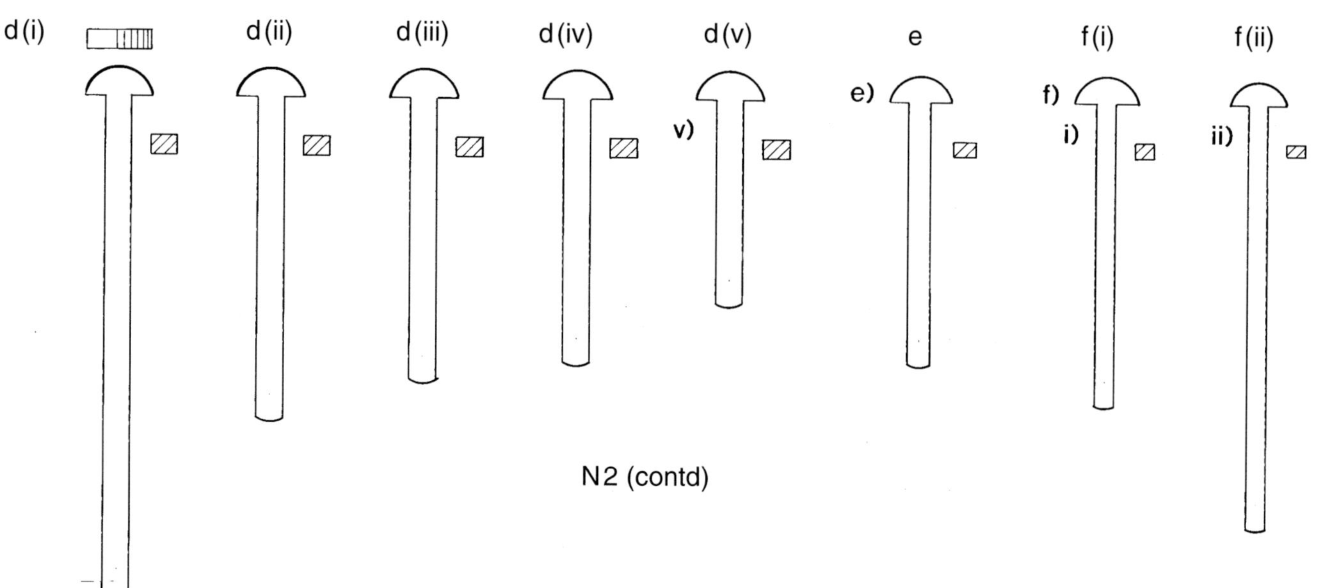

d(i)
d(ii)
d(iii)
d(iv)
d(v)
v)
e
e)
f(i)
f)
i)
f(ii)
ii)
N2 (contd)
N3 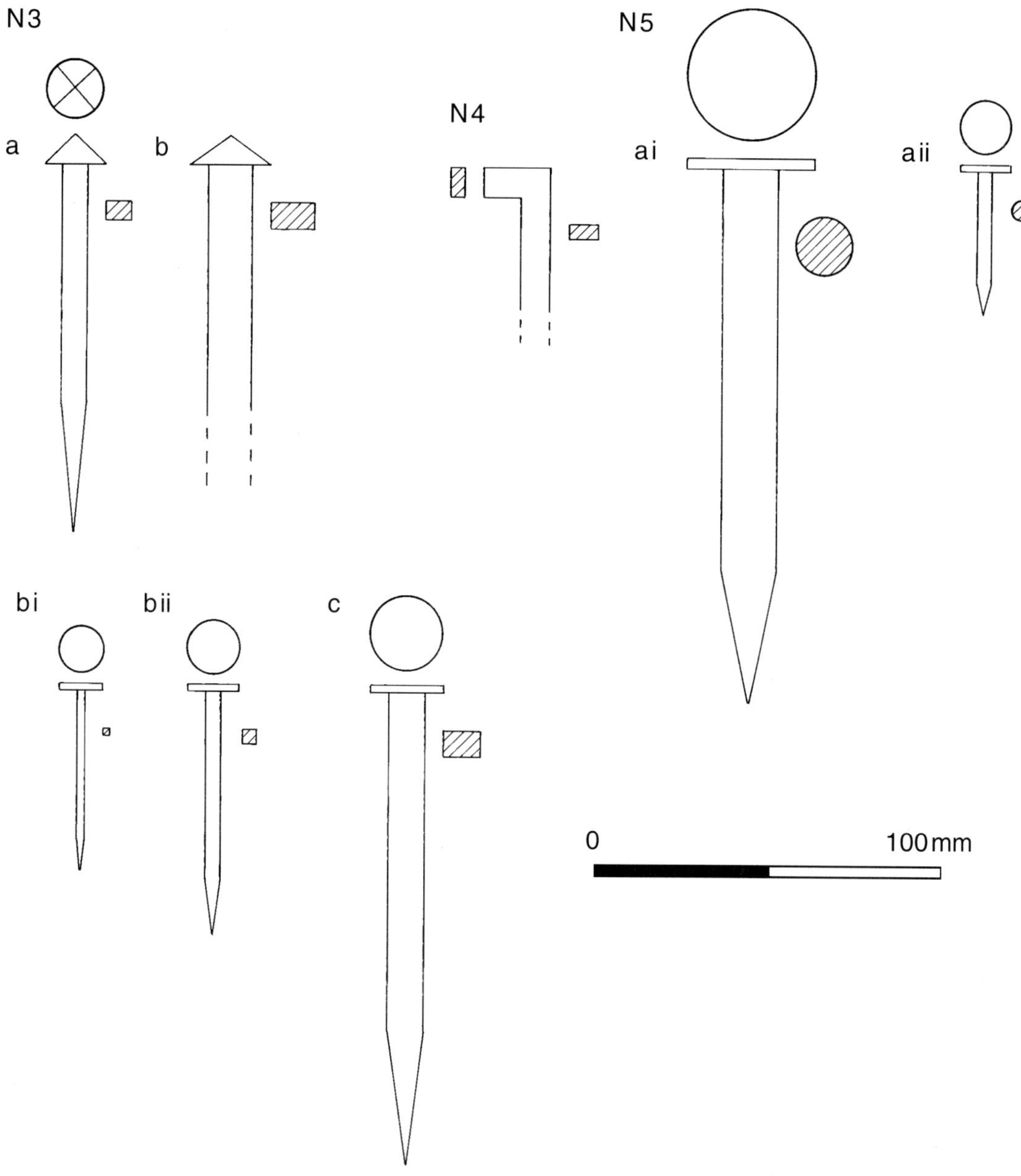
N4
N5
a
b
ai
aii
bi
bii
c
0
100mm

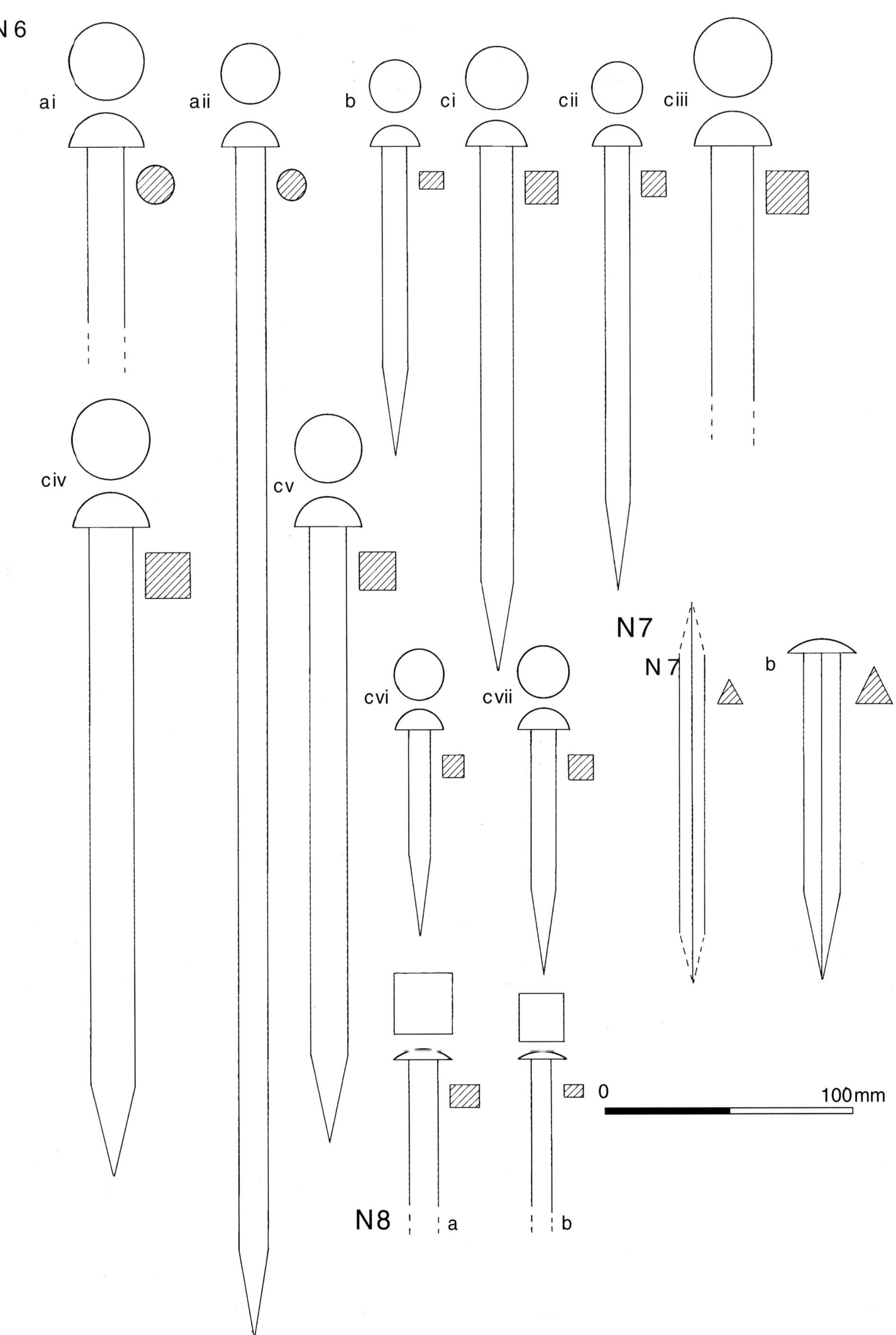

Fig 21   Ironwork: nail types N6 – N8

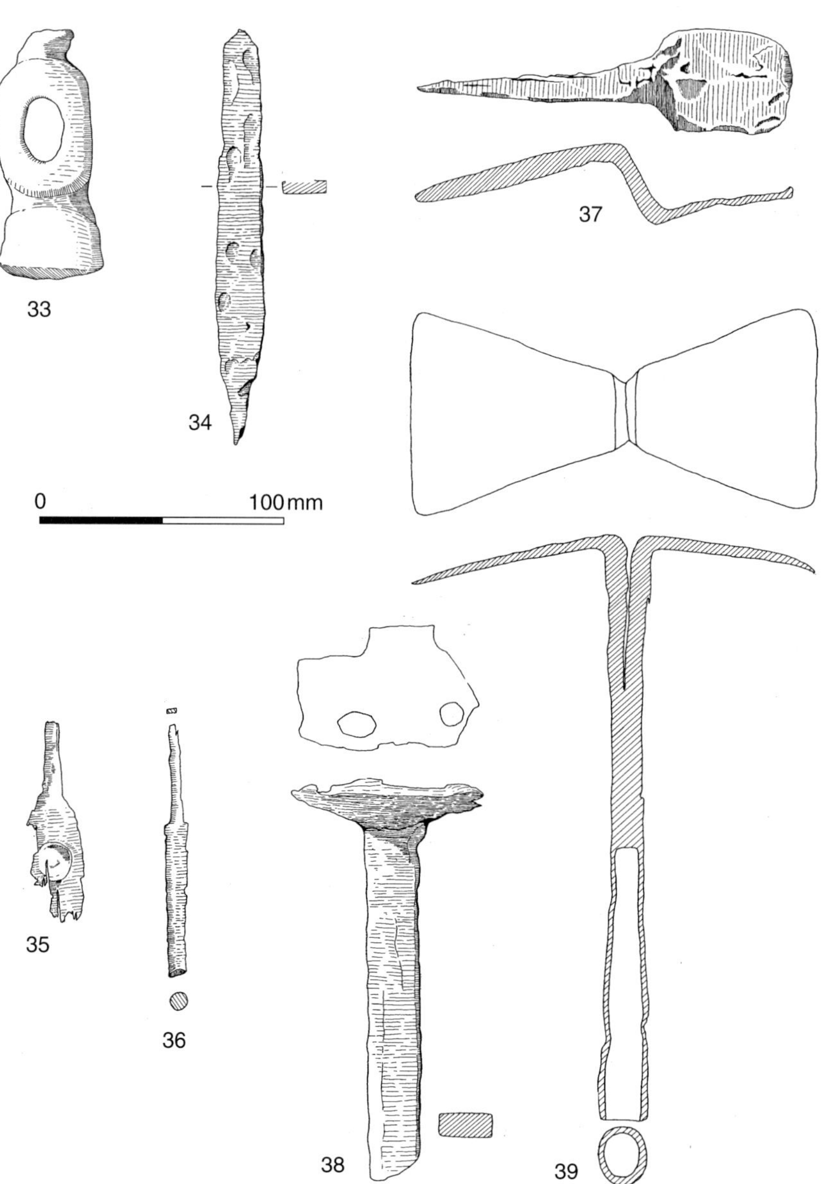

Fig 22 Ironwork: tools

retained; however all firebrick from earlier contexts was retained.

Neither the technology nor the chronology of firebrick development have been adequately studied; as well as form and fabric, the stamps found on some examples should in principle provide valuable dating evidence. The available information on producing firms for the region is contained in Davison 1986; Gurke (1987) reviews the use of stamped firebrick in an export context. It has not proved possible to fully investigate these aspects of the Derwentcote assemblage, which would repay further study.

The fabrics could not be fully classified by visual examination. Almost all were coarse, and were tempered with grit, coarse grit, and very coarse grit; other inclusions included pebbles and grog. Some had probably been affected by prolonged heat during use, producing a 'stoneware-like' appearance merging into vitrification.

The range of forms (Fig 24, and Davison 1986, 219) was quite wide; dimensions show some variation around the averages quoted (in Imperial, as being the system to which the bricks were manufactured). Square bricks were the commonest (248 examples, excluding stamped bricks), but there were also appreciable numbers of plain quarls (66), splits (37), bull-heads (19), side-wedges (12), and plinths (11), remaining types having less than ten examples. A few specimens showed combination shapes (*eg* bull-heads with end-wedge or feather, bullhead with rebate, end- and side-wedge). In addition, there were two plain quarls with grooves along their beds, and one square brick measuring 4′ x 4′ x 2″.

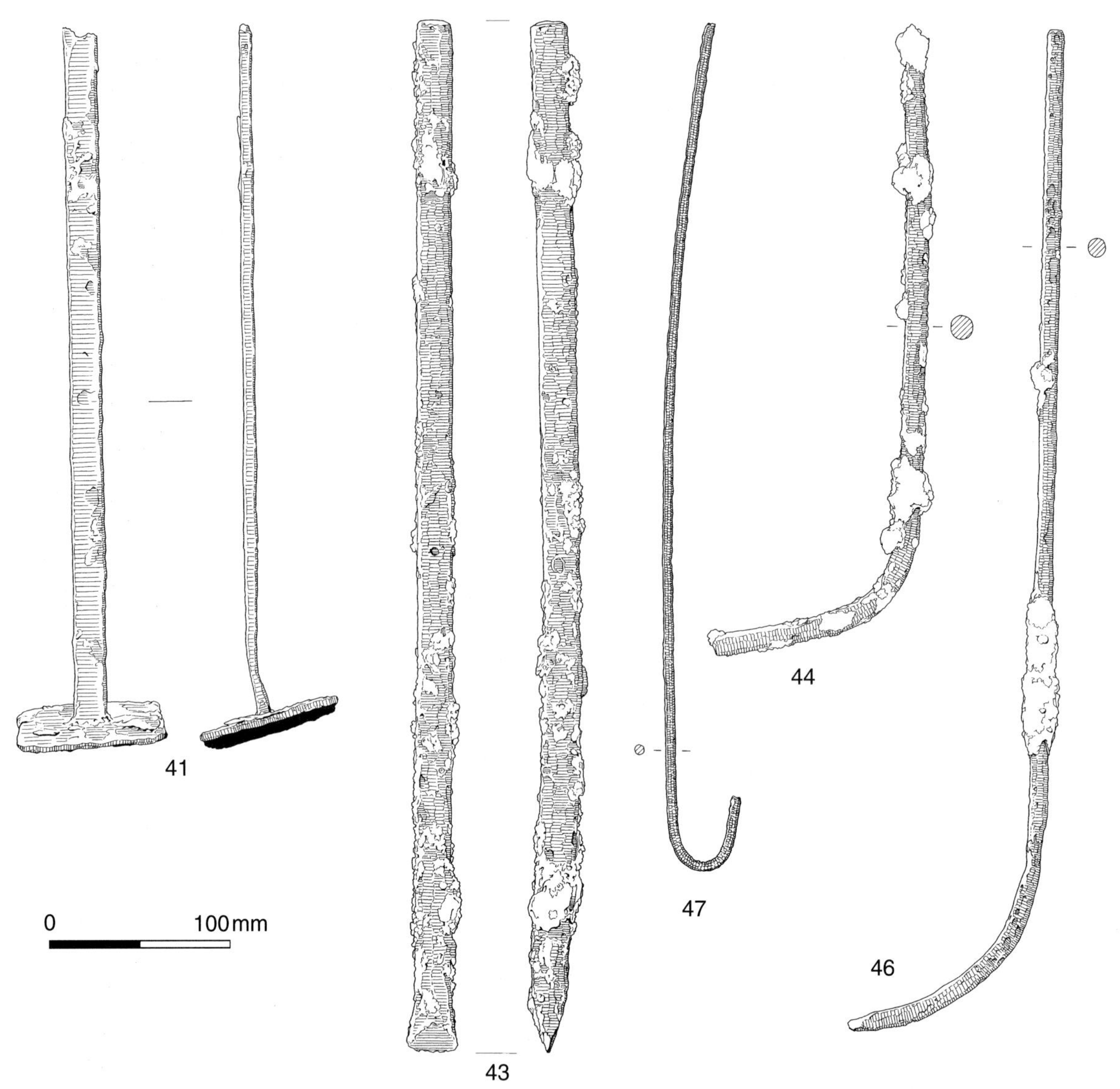

Fig 23 Ironwork: chisel (43), and unidentified objects

The proportion of stamped bricks was quite high (almost one-fifth of the recovered assemblage, though this will have been biassed slightly by the retention policy). In addition to the finds assemblage, stamped firebricks were also recorded in the agricultural period floors of the Northern and Southern Buildings (stamps within the *in situ* structure were not normally visible, due to their location on the bed faces of the bricks, and to vitrification of most exposed surfaces). The following stamps were present:

BUTE. 20 examples, including four closers, almost all from infills of the southern ashpit. The Bute Pit firebrick works at High Spen (3.5 km north of Derwentcote) operated from 1875 to 1958 (Davison 1986, 171, 175).

R. DICKINSON & CO CONSETT NEWCASTLE on TYNE. The stamp is an oval, with 'CONSETT' across the centre. 20 examples, all certain or probable square bricks. One example from a late (but probably pre-closure) raft of rubble below a loading hole of the furnace (*Ch 6*); remainder from post-closure contexts in various areas, including nine from recent collapse rubble over the northern ashpit. R Dickinson & Co operated Dickinsons Yard, Consett (6 km south of Derwentcote), from 1855 to 1879 (Davison 1986, 170, 171).

DODD [?]. One example, on heavily vitrified brick from agricultural conversion of Southern Building. Perhaps M DODD of Dilston Park, Corbridge (17 km west of Derwentcote), producing stamped bricks in the 1870s (Davison 1986, 86–7, 112).

HAMST.ERLEY. 14 examples, all square bricks where measurable, from post-closure contexts mainly in the Northern Building. The Hamsterley Colliery brickworks (1 km west of Derwentcote) was active 1874–1924 (Davison 1986, 170–171).

HANNINGTON. Three square bricks, from post-closure contexts. Hannington & Co produced fire-bricks at Swalwell (9 km north-east of Derwentcote) from 1850 to 1906 (Davison 1986, 131).

LINTZ. 34 examples, including two side-wedges and one closer, showing three different stamps (plain LINTZ, .LINTZ., and LINTZ plus thumbprint); three specimens stratified within the industrial sequence of the Southern Building, the remainder from post-closure contexts, including several derived from collapse of brick patchings to the Eastern Building. The Lintz Colliery brickworks (3 km east of Derwentcote) operated from either 1864 to 1880, or 1868 to 1924 (Davison 1986, 170, 171).

RAMSAY and [G] H RAMSAY NEWCASTLE. 107 (including seven bull-head) and 12 examples respectively; by far the commonest manufacturer. George Hepple Ramsay (grandson of one of the Hepples in the Derwentcote workforce(*Ch 4*)) and his descendents operated brickworks under the G H Ramsay name at Derwenthaugh and Swalwell (both 9 km north-east of Derwentcote), from 1810 to 1880 and 1830 to 1920 respectively (Davison 1986, 132, 13–8, 279); G H Ramsay was a partner in Derwentcote from 1873 to 1875 (*Ch 4*). Both stamps occurred in a range of post-closure contexts, with large quantities in the northern ashpit and the cone interior (in both of which RAMSAY firebricks were visible *in situ* in the furnace structure), and also in the late (but probably pre-closure) rubble banks against the west face of the furnace. It is clear that 'Ramsay' bricks were the main supply used for lining and repairing the furnace in its final years, if not before.

RITSON (with the 'N' stamped backwards on some examples). 27 specimens, including seven square bricks and four identifiable bull-heads. Most examples were from the rubble bank against the west face of the furnace, and from the agricultural conversion and topsoil of the Southern Building; this suggests use in the chest compartment, rather than in the firegrate area. Presumably from H Ritson's Carr House firebrick works, Consett (6 km south of Derwentcote), active in 1879 (Davison 1986, 170, 281).

V.G.C. Ten examples, including one side wedge; one from the final metalling of the trackway in the Northeast exterior, the remainder from post-closure contexts in various areas. Produced at Victoria Garesfield Colliery brickworks (3 km north-east of Derwentcote), active from 1875 to at least 1914 (Davison 1986, 170–1, 281).

WILKINSON [?]. Three stamps, all indistinct; it is possible that they in fact read 'DICKINSON', or that they are the products of W Wilkinson, active at Prudhoe in 1902 (Davison 1986, 87).

In addition to the name stamps, and the 'frogged' firebricks within the non-stamped assemblage, a group of 13 firebricks contain a distinctive combination of a 'frog' with a deep thumbprint-like indentation; this is sufficiently consistent to appear deliberate, rather than being merely the result of pushing clay out of a sticky mould. These came from post-closure contexts, with the exception of one possibly-earlier context in the Southern Building.

In assessing the dating evidence of the stamps, it must be remembered that the large majority are from post-closure contexts; while many of these will have derived from the furnace, some may have been introduced after closure, either from the forge or from the manufacturer. As noted above, only RAMSAY firebricks could actually be identified *in situ* in the structure, though the distribution of BUTE, DICKINSON, LINTZ, and RITSON does positively suggest derivation from the structure. The presence of special forms in VGC also suggests industrial use. The firebrick evidence therefore suggests that the furnace was still being used and repaired after 1875, though the identification of *in situ* examples would be needed to confirm this.

Stone

One piece of moulded mullion of Victorian style was recovered from the ashpit. Like the decorated glass, this was out of character to the utilitarian and vernacular construction of all surviving features of the furnace, and is likely to have originated from Derwentcote House.

In addition, large amounts of white sandstone, and smaller amounts of red sandstone, were retrieved. The former definitely, and the latter probably, were used as refractory contructional materials within the furnace, and they are therefore classified as process residues.

Process residues

Unsurprisingly, a high proportion of the total 'finds' assemblage consisted of materials used or formed in the operation of the furnace. This classification to some extent crosscuts the categories of 'artefact' and 'building material' already discussed. For practical purposes, the ironwork assemblage has been

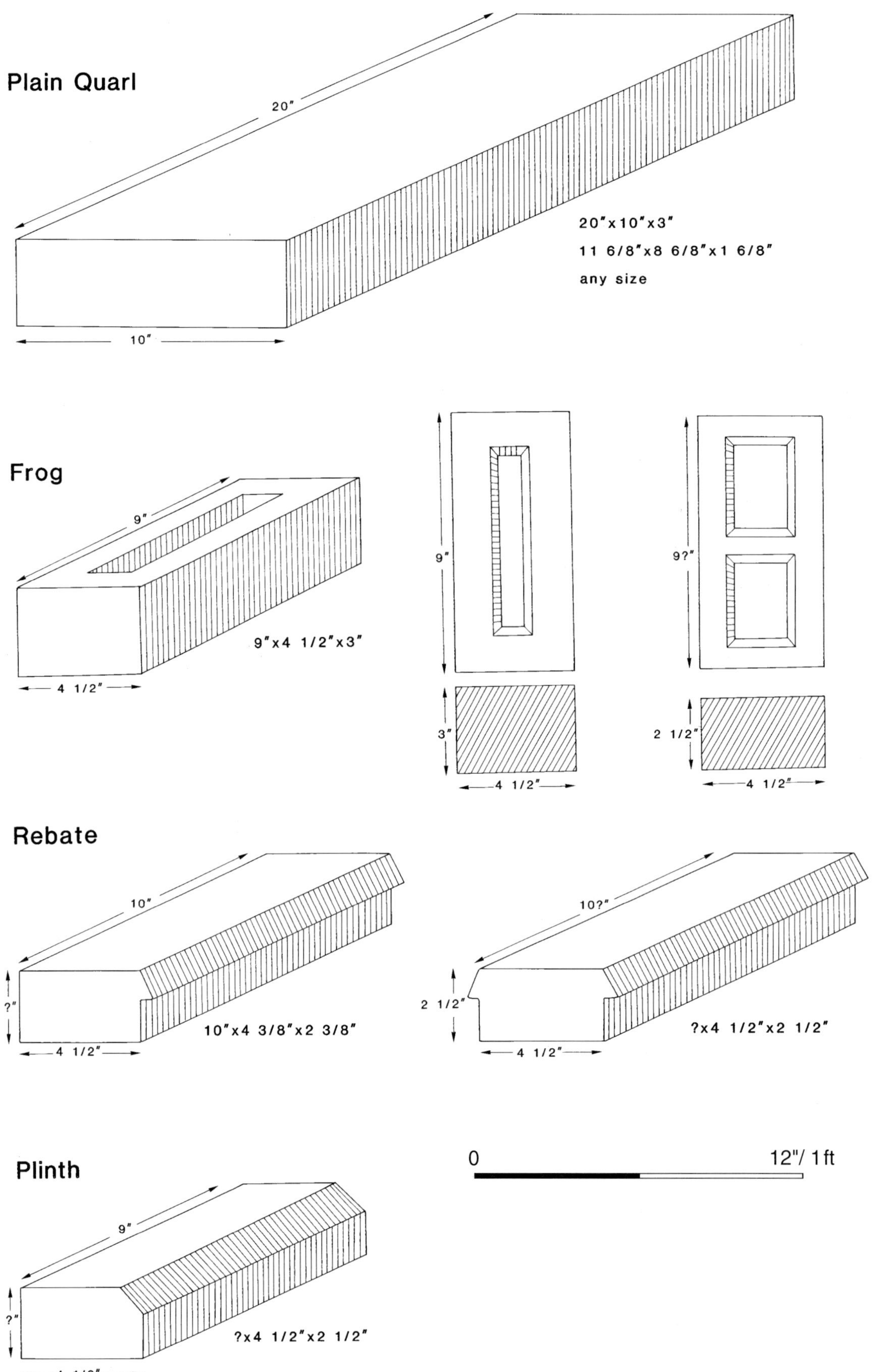

Fig 24 Firebricks: form series

classified under artefacts, and firebrick under building materials. Remaining refractory materials and raw materials are however considered as process residues, along with slags and those samples of iron selected for metallographic examination. The bulk of these materials were classified as 'slag types' A–M (of which Types K and L were later abandoned). These are the subject of McDonnell's technical report below; remaining materials are briefly listed and described.

Red sandstone

Appreciable quantities of red sandstone, sometimes identifiable as fragments of squared blocks, were recovered; in notable contrast to most other building materials, the bulk of this material came from stratigraphically-early contexts, in the north-east exterior and the Southern Building. One piece was vitrified on one face, and vitrified red sandstone was also present in patching of the Southern Building east wall (*Ch 5*). The stone is unlikely to be of local origin, and may well have been imported to the site as refractory lining material, perhaps in earlier phases before the widespread adoption of firebrick.

The slags and residues
Dr Gerry McDonnell, site information added by D Cranstone

Introduction and methodology

The technology of cementation steelmaking has been described by Cranstone above (*Ch 2*). The raw materials for the process were high-grade bar iron, charcoal (or other carbon-based carburising mixtures), coal fuel, and the various refractory materials used to build, lute, and seal the furnace and cementation chests; the products were the bars of steel, unconsumed carburising material, chest sealing material, the chests themselves, and ash and cinders from the furnace. It should be noted that no silicate slag was generated as a byproduct. However it must be remembered that a wider range of processes (including finery and puddling production of wrought iron, crucible steelmaking, and both forging and rolling of iron and steel) was conducted at Derwentcote Forge, and residues from these processes may be included in the assemblage from the furnace site (as may extraneous material brought in as (for example) road metalling). Indeed the assemblage includes at least one material (crucible pot-lids, Slag Type H) demonstrably derived from the forge.

A reference set of the different slags and residues was established during the excavation of the site. These are referred to below as 'Slag Types', though they include other process residues as well as true slags. In general the residues were readily ascribed to one of the types, with few cases of significant doubt. This does imply that subtle differences within residue type were not identified. In general the types fell into three groups: true slags, metallic waste, and refractory material.

The slags and residues were morphologically described and samples were then analysed. Reflected light microscopy of polished thick sections was used to determine the mineral texture. Scanning electron microscopy with an attached energy dispersive X-ray analysis system was used to obtain bulk area and phase analyses (results are given as oxide weight percent). X-Ray Diffraction was used to identify the minerals present in some samples.

Slag Type A

Morphology
Irregular-shaped lumps ranging in size from tens of millimetres to *c* 300 mm, and in weight from tens to hundreds of grams. It has a high "apparent density" (*ie* . it contains few vesicles), and a fine crystalline fracture; in one direction (vertical?) the crystals are columnar. Some surfaces show characteristic flow effects *eg*. ripples, and it is clear from the morphology that the slag had been fully liquid, and had cooled very quickly. The fractured surface is usually grey and often has an almost metallic lustre, but in a cut cross-section it is grey/green in colour.

Mineral texture
Fine dendritic silicate laths and finely distributed metallic inclusions in a glassy matrix (Plate 46).

Chemical and mineral composition
The area and phase analyses are given in Table 1. The area analyses exceed 100% due to the presence of the metallic inclusions. The area bulk analyses (B1–B5) show an essentially fayalitic composition but with a high alumina content. The analysis of the silicate phase confirms it as fayalite ($2FeO.SiO_2$). The composition of the glassy phase is a silica-aluminium oxide rich phase, broadly an alumino-silicate. A normative calculation carried out on this composition gives the predominant minerals as quartz (SiO_2, 23.1%), ferrosilite ($2FeO.Si_2O_3$, 21.1%), anorthite ($CaO.Al_2O_3.2SiO_2$, 19.5%), hercynite ($FeO.Al_2O_3$, 11.4%) and orthoclase ($K.AlO_2.3SiO_2$, 10.9%). This shows that the glassy phase is saturated in silica. The analysis of the crystal (XTAL, Table 1) is an iron oxide phase rich in alumina (Al_2O_3) and titania (TiO_2) which is indicative of a spinel structure, of magnetite/ulvite ($Fe_3O_4/2FeO.TiO_2$) and hercynite ($FeO.Al_2O_3$).

93

	B1	B2	B3	B4	B5	SIL	GLAS	XTAL
Na_2O	0.7	0.7	0.5	n.d	0.6	n.d	0.8	0.2
MgO	0.5	0.5	0.3	0.2	0.4	0.1	0.1	0.1
Al_2O_3	10.3	8.4	8.5	9.3	9.0	0.1	17.6	11.7
SiO_2	36.9	36.6	38.5	39.3	39.3	31.1	54.7	2.1
P_2O_5	0.7	0.6	0.4	0.6	0.4	n.d	1.1	0.0
S	0.4	0.4	0.4	0.3	0.4	n.d	1.3	0.0
K_2O	1.0	0.8	0.9	0.9	1.1	0.1	1.9	0.0
CaO	1.4	1.3	1.6	1.6	1.4	0.1	5.5	0.1
TiO_2	0.3	0.2	0.4	0.3	0.3	n.d	0.3	2.1
Cr_2O_3	0.2	n.d	n.d	n.d	0.1	0.2	n.d	0.2
MnO	0.1	0.4	0.3	0.2	0.3	0.5	0.3	2.1
FeO	54.3	52.1	50.9	52.4	53.0	69.7	20.1	79.3
CoO	0.1	n.d	n.d	n.d	n.d	n.d	n.d	0.0
NiO	n.d	n.d	n.d	n.d	n.d	n.d	n.d	0.0
CuO	n.d	0.2	0.2	0.1	n.d	0.1	0.1	0.1
PbO	n.d	n.d	n.d	n.d	n.d	n.d	n.d	0.3
	106.9	102.2	102.9	105.2	106.3	102.0	103.8	96.2

Table 1   Analyses of Slag Type A

Discussion and interpretation

The morphology and mineral texture indicate that the slag cooled very rapidly from a high temperature. The chemical composition confirms this since the glass is saturated in silica which had insufficient time to crystallise. The high silica and alumina contents are indicative of an iron/slag/lining reaction.

The analysis shows that Slag Type A is a free-running iron silicate slag. Such slags can be generated by a number of different processes, of which two are the most likely genesis in this case. This slag type could either be a product of the direct reduction process (*eg* the bloomery process, or a variant) or a finery or puddling slag, *ie* the refining of cast iron into wrought iron. The high alumina content is not typical of bloomery slags (normally *c* 5%), and there are very few analyses of finery or puddling slags. Killick and Gordon (1987) published the results of a microscopy analysis of some puddling slags but only included phase analyses. Percy (1864, 668) also gave some analyses of puddling "tap cinder", all of which contained less than 3% alumina. Tylecote (1986, 219) gives two analyses of finery slags which are similar in composition to the direct reduction slags. Therefore the Derwentcote Type A Slag cannot be reliably attributed on analytic grounds

The Type A assemblage at Derwentcote is extremely large (over 520 kg). The bulk of this assemblage came from late or post-abandonment contexts, including trackway surfaces where it had been used as metalling. However there were smaller but still appreciable quantities stratified within earlier contexts. It is therefore likely that Slag Type A derived from either a long-lived process, or is the conflated product of several different processes. Since it cannot plausibly be derived from the cementation process, it is likely to have been imported to the furnace area, probably from the forge. An origin in one or more of the iron-refining processes (fining or puddling) is tentatively suggested.

Slag Type B

Morphology

A reddish/black slag of varying texture. Overall it has a cindery appearance and a low "apparent" density. It occurs in plates between about 40 and 10 mm in thickness, and up to 100 mm maximum breadth. Although many pieces had edges, either slightly curved or straight, no complete pieces were recovered. It was not possible to determine which way-up the slag formed, although there was clear evidence in some cases for a flat surface formed by coal or charcoal. The 'flatness' of the slag suggested that it formed on the surface of a chest or crucible.

The slag had a cindery brittle fracture, and the fractured surface showed that it was heavily vesicular. The cut cross-section showed a cindery and vesicular slag with some white (silica or ash?) inclusions and one charcoal/coal lump.

Mineral texture
Under reflected light the slag comprised mostly (95%) a single phase, but with an occasional angular phase present. The slag was heavily vesicular.

Chemical and mineral analysis
Bulk analyses B1 and B2 (Table 2) were obtained from areas close to the coal, and were therefore probably not simple oxides, hence the low totals. Analyses B3–B5 were obtained from the 'denser' slag areas. All analyses indicate alumino-silicate compositions but with a high titania level. This is due to the presence of titania-rich inclusions (Incl 1) which were surrounded by a silica film. A fine crystallite was also analysed (XTAL, Table 2) and shown to be rich in silica, alumina and iron. X-Ray Diffraction (XRD) analysis identified the presence of mullite.

Discussion and interpretation
The morphology indicated an agglomerated type of slag formed by solid state reactions, rather than minerals crystallising from an alumino-silicate melt. The morphology, texture and analysis show that the slag formed as a result of a high-temperature reaction involving a furnace or hearth lining (perhaps similar in composition to Type D slag), the predominant phase being mullite. The high titania values are of particular interest.

Slag Type C
Morphology
A friable fritted clay and sand mix with a red body and a black surface. The maximum surviving thickness was about 35 mm. The black layer varied in thickness with a maximum of 15 mm, and formed a reasonably flat surface, but it was not clear which (red or black layer) was the external surface.

Mineral analysis
Samples of the red and black layer were analysed by XRD, but only the presence of silica was confirmed. Other lines were present but could not be ascribed to minerals. Qualitative X-Ray Fluorescence (XRF) analysis detected the presence of iron and silicon with minor levels of aluminium, potassium, calcium, and titanium present.

Discussion and interpretation
Slag Type C is essentially a sand layer that had been reduced (black) or oxidised (red) and was fritted together. It is interpreted as used sand capping from the cementation chests (*Ch 2*), the oxidised (red) sand being the external surface and the reduced (black) layer being the internal surface over the bars. The reducing atmosphere penetrated the capping sand. Examination of the *in situ* chests showed that the top 100–200 mm of the chest walls were oxidised, indicating a minimum thickness for the capping.

	B1	B2	B3	B4	B5	Incl1	XTAL
Na_2O	0.3	n.d	0.8	0.1	0.2	0.2	0.2
MgO	0.7	0.6	0.5	0.5	0.7	0.5	0.7
Al_2O_3	21.2	20.0	23.7	26.7	16.9	3.2	12.4
SiO_2	43.1	40.2	56.4	56.9	62.1	16.5	58.5
P_2O_5	0.2	0.1	0.6	0.5	0.4	0.6	0.3
S	0.1	0.2	0.1	0.1	n.d	0.2	n.d
K_2O	2.1	1.9	1.3	1.3	1.3	0.9	1.2
CaO	0.13	0.3	1.4	1.9	2.1	0.2	1.8
TiO_2	1.0	1.0	1.3	1.3	1.6	88.6	2.5
Cr_2O_3	n.d	n.d	n.d	n.d	n.d	0.2	n.d
MnO	0.1	n.d	n.d	n.d	n.d	0.1	n.d
FeO	1.3	1.8	9.6	9.4	12.1	1.5	25.9
CoO	0.1	n.d	n.d	n.d	0.1	n.d	n.d
NiO	n.d	n.d	n.d	0.2	n.d	n.d	0.1
CuO	0.1	n.d	n.d	0.1	0.1	n.d	n.d
PbO	n.d	n.d	n.d	n.d	n.d	n.d	0.1
	70.6	66.1	95.7	99.0	98.0	112.7	103.7

Table 2   Bulk and phase analyses of Slag Type B

Slag Type D

Morphology

Slag Type D was a dense vitrified brick, probably highly refractory. Some of the fragments were heavily slagged. A sample with slagging was selected for analysis.

	B1	B2	B3
Na_2O	0.4	0.4	0.2
MgO	1.0	0.7	0.9
Al_2O_3	29.8	28.0	31.7
SiO_2	56.4	65.6	50.8
P_2O_5	n.d	0.1	0.2
S	n.d	n.d	n.d
K_2O	1.8	2.2	1.2
CaO	0.3	0.6	2.2
TiO_2	1.5	1.6	1.4
Cr_2O_3	0.1	n.d	n.d
MnO	n.d	n.d	0.1
FeO	2.2	2.6	4.7
CoO	n.d	0.1	n.d
NiO	0.1	n.d	0.1
CuO	0.1	n.d	n.d
PbO	n.d	0.2	n.d
	93.7	102.1	93.5

Table 3   Bulk area and phase analyses of Slag Type D

Mineral texture

After polishing the darker slagged zone could be clearly distinguished from the light grey unreacted brick. However, under reflected light the two zones were very similar. The unreacted brick comprised a pale grey matrix with vesicles and very fine small spheroidal highly reflective inclusions. The slagged zone displayed a similar structure but with a greater number of vesicles and inclusions.

Chemical and mineral analysis

The analyses (Table 3) show the brick to be aluminosilicates. XRD analysis confirmed the presence of mullite.

Discussion and interpretation

The analytical evidence supports the initial interpretation that Slag Type D is refractory brick which on some faces has suffered slag-attack.

Slag Type E

Morphology

The Type E slag occurs in small irregularly-shaped pieces up to 50 mm maximum diameter. It has a glassy appearance and a low density (relative to silicate slags). The surface is usually blue/black in colour, and it has a conchoidal fracture, the surface of which is green. When sectioned it shows few vesicles and is blue/grey in colour.

Mineral texture

Examination showed that it had no crystalline structure, but consisted of a glassy matrix with prills of metallic iron.

Chemical and mineral analyses

The bulk analyses (B1–B3, Table 4) show that the slag is rich in silica and alumina with minor levels of magnesia, potash, lime, and iron oxide. Manganese and titanium oxides are also present. Analysis of the metallic inclusion showed that it was a cast iron, containing about 3% carbon and a significant phosphorus content. XRD analysis confirmed the presence of mullite, indicating that some crystallisation had occurred.

Discussion and interpretation

The colour morphology, absence of a mineral microstructure, and chemical composition are typical of a blast furnace slag. A comparative analysis of blast furnace slag from the seventeenth-century furnace at Allensford (8 km south-west of Derwentcote) is also given in Table 4 (Linsley and Hetherington 1978; spectrographic and chemical analyses). The Derwentcote slag is significantly higher in magnesia than the Allensford material, and is therefore unlikely to come from that site.

The total quantity of Type E slag was small (just under 1 kg), but the material was widely distributed on site, and a relatively high proportion came from early contexts (including fossil soils cut and sealed by the construction of the furnace); given the small volumes of such contexts that were excavated, it appears that Type E slag is concentrated in early contexts, and may well be residual in later contexts. The small quantity could conceivably have been introduced as encrustations on bars of pig iron for the forge (bar iron for cementation should not contain any trace of blast furnace slag), or as road metalling (though this would imply a furnace reasonably close to Derwentcote); alternatively, it is conceivable that they derive from an undocumented blast furnace at Derwentcote itself (for which the forge area would be topographically suitable).

Slag Type F

The Type F slag was corroded iron or steel. An example was metallurgically examined and shown to comprise pearlite plus grain-boundary ferrite, with a carbon content of $c\,0.7–0.8\%$.

	B1	B2	B3	Prill		Allensford Slag	
Na_2O	0.5	0.6	0.3	Na	n.d	Na_2O	1.79
MgO	7.5	7.7	7.4	Mg	n.d	MgO	2.65
Al_2O_3	18.8	19.4	20.0	Al	n.d	Al_2O_3	20.33
SiO_2	51.3	51.1	52.8	Si	0.2	SiO_2	48.25
P_2O_5	0.2	0.1	0.1	P	0.7	P_2O_5	n.d
S	0.1	n.d	n.d	S	0.1	S	0.1
K_2O	3.2	3.0	3.1	K	n.d	K_2O	3.09
CaO	7.3	7.2	7.5	Ca	n.d	CaO	6.81
TiO_2	0.9	1.0	0.9	Ti	n.d	TiO_2	n.d
Cr_2O_3	n.d	n.d	n.d	Cr	0.1	Cr_2O_3	n.d
MnO	2.3	2.2	2.3	Mn	n.d	MnO	1.38
FeO	6.6	6.4	5.8	Fe	95.2	FeO	9.71
CoO	n.d	0.1	n.d	Co	0.1	CoO	n.d
NiO	n.d	0.1	n.d	Ni	n.d	NiO	n.d
CuO	n.d	n.d	0.1	Cu	0.2	CuO	n.d
PbO	n.d	n.d	0.2	Pb	n.d	PbO	n.d
	98.7	98.9	99.7	96.6		94.01	

Table 4 Slag Type E area bulk analyses and analysis of a metallic prill. Analysis of Allensford charcoal blast furnace slag for comparison

Slag Type G

Morphology

Slag Type G occurs as irregularly shaped lumps, up to 100 mm maximum length and usually between 50 and 100 mm in thickness. External surfaces range between agglomerated textures and vitrified surfaces. The surface colour is black but there are white inclusions (coal or quartz?) entrapped throughout the structure. In cross-section the slag is heavily vesicular, black in colour but showing the numerous white inclusions (which range in size up to 5 mm across).

Mineral texture

Fine dendritic (iron?) oxides and massive silicate(?) with a very small amount of a third phase present. In the prepared section there were no white inclusions.

Chemical and mineral composition

The bulk area analyses (Table 5) show that the slag is rich in silica, alumina and titania but low in potash, lime and iron oxide. The phase analyses (Table 5a) show that the massive phase is an alumino-silicate, the dendritic phase is an ulvo-spinel, and the third phase is alumina-rich. The phase analyses have low totals, particularly the third phase, indicative of higher oxidation states. The normative calculations give the major phase as hercynite and silica-rich minerals, *eg.* orthoclase and quartz. The massive phase and the third phase are also silica-saturated; excess silica (calculated as quartz) is present in both phases. The other major calculated phases in the massive phase are hercynite, orthoclase and anorthite. Those in the third phase are mullite and hercynite. The dendritic phase is a spinel, the major calculated phases being hercynite and wustite. The theoretical calculations determined that quartz was a constituent of the slag, but it was not observed either optically or in the SEM, indicating that the slag deviated strongly from equilibrium conditions. XRD analysis only detected the presence of mullite.

A metallic prill was analysed and shown to be lead with 1% aluminium present.

Discussion and interpretation

The slag is a high silica-alumina slag which has deviated strongly from equilibrium conditions. The slag morphology and mineral texture show that the deviation was not due to rapid cooling from a fully liquid state. The slag was probably formed in the solid state under relatively oxidising conditions in a silica-alumina rich environment. Such conditions could prevail in some hearths or stoke holes. The presence of possible coal would support both hypotheses.

The presence of a lead prill within the structure is unexpected. It is surprising that it survived in the metallic state rather than being oxidised, given the overall oxidation level of the slag.

Slag Type H

Morphology

Circular discs of dense refractory material, c 150 mm in diameter and 20–30 mm thick. The upper and lower surfaces were flat and vitrified, although one side was 'reflective' and the other quite dull. The

	B1	B2	B3	B4	B5
Na₂O	1.0	0.1	0.2	0.3	0.6
MgO	0.9	0.7	0.4	0.5	0.6
Al₂O₃	27.7	25.3	22.3	31.5	24.6
SiO₂	47.5	50.8	50.2	46.6	48.2
P₂O₅	0.2	0.4	n.d	0.2	0.3
S	0.1	n.d	n.d	n.d	n.d
K₂O	1.6	1.9	1.9	1.5	1.5
CaO	1.7	0.8	1.5	0.9	1.4
TiO₂	1.3	1.6	1.5	1.0	1.4
Cr₂O₃	n.d	0.1	n.d	n.d	0.1
MnO	n.d	n.d	n.d	n.d	n.d
FeO	20.0	14.2	20.0	14.7	21.3
CoO	n.d	n.d	0.2	n.d	0.1
NiO	n.d	n.d	0.2	n.d	n.d
CuO	0.2	n.d	n.d	n.d	n.d
PbO	n.d	0.3	0.2	n.d	n.d
	102.2	96.2	98.6	97.2	100.1

Table 5   Slag Type G bulk area analyses

vitrification extended around the side of the disc. The vitrification was brown/black in colour. In section the discs had a uniform texture, except that at the reflective surface there was a thin white band. The remainder was a typical dense refractory material.

Mineral texture
Only one major phase could be distinguished using reflected light microscopy, but there were some small round metallic(?) inclusions present.

Bulk and phase analysis
The bulk and matrix analyses (Table 6) show that the material is a silica rich alumino-silicate. The metallic prill is a cast iron with a high silica content. XRD analysis showed the presence of mullite.

Discussion and interpretation
Refractory lids for sealing crucible steel melting pots.

	Massive Phase	Dendritic Oxide	Third Phase		Lead Inclusion
Na₂O	1.0	0.6	n.d	Na	0.2
MgO	0.7	1.5	0.4	Mg	0.1
Al₂O₃	15.9	24.1	57.7	Al	1.2
SiO₂	57.7	4.4	24.0	Si	0.7
P₂O₅	0.2	n.d	n.d	P	n.d
S	n.d	n.d	n.d	S	0.1
K₂O	1.9	0.1	n.d	K	n.d
CaO	2.3	0.1	n.d	Ca	n.d
TiO₂	2.0	1.7	0.4	Ti	n.d
Cr₂O₃	n.d	n.d	0.3	Cr	0.1
MnO	0.1	n.d	n.d	Mn	0.1
FeO	12.2	61.4	4.7	Fe	0.6
CoO	n.d	n.d	n.d	Co	n.d
NiO	0.1	n.d	n.d	Ni	n.d
CuO	n.d	n.d	n.d	Cu	0.1
PbO	n.d	0.1	0.1	Pb	96.5
	93.2	94.0	87.6		99.8

Table 5a   Slag Type G phase analyses

	Bulk1	Bulk2	Matrix		Prill
Na_2O	0.1	0.4	0.2	Na	0.5
MgO	0.8	1.1	1.5	Mg	n.d
Al_2O_3	31.9	32.6	38.5	Al	0.1
SiO_2	51.8	53.5	59.7	Si	10.2
P_2O_5	0.2	0.2	0.4	P	0.2
S	0.5	0.4	0.7	S	0.2
K_2O	1.7	1.9	2.0	K	n.d
CaO	0.2	0.3	0.4	Ca	n.d
TiO_2	1.5	2.0	1.2	Ti	1.0
Cr_2O_3	0.1	n.d	n.d	Cr	0.1
MnO	0.1	n.d	n.d	Mn	n.d
FeO	3.8	2.4	1.0	Fe	92.1
CoO	n.d	n.d	n.d	Co	n.d
NiO	0.1	n.d	n.d	Ni	0.4
CuO	0.1	0.1	n.d	Cu	n.d
PbO	0.2	n.d	n.d	Pb	n.d
	93.1	94.9	105.6		103.9

Table 6   Slag Type H, bulk, matrix, and prill analyses

Slag Type I

This slag type was initially established since the first samples appeared sufficiently distinct to justify another Type, however subsequent analysis has shown that they were slag-attacked or vitrified examples of Type D, refractory brick. The analyses (Table 7) are similar to those obtained for Type D, but the presence of a copper prill is unexpected.

Slag Type J

A friable ferruginous concretion with black manganese oxide inclusions. The material is of natural origin; it may have formed naturally on site, or have been introduced.

Slag Type K

Field category, abandoned on further analysis.

Slag Type L

Field category, abandoned on further analysis.

Slag Type M

Morphology

This slag type has a similar morphology to classic iron smelting tap slag. The upper surface has a ropy flowed appearance, with large vesicles. The slag occurs as plates, c 20–50 mm thick, and up to about 100 mm length and/or breadth.

	B1	B2	B3		Prill
Na_2O	0.5	0.7	n.d	Na	n.d
MgO	1.2	0.9	0.7	Mg	0.4
Al_2O_3	31.2	28.7	26.1	Al	0.6
SiO_2	56.8	57.6	53.9	Si	0.5
P_2O_5	0.2	0.3	0.2	P	n.d
S	0.1	0.2	0.1	S	n.d
K_2O	2.0	2.1	0.3	K	n.d
CaO	0.3	0.4	0.3	Ca	0.1
TiO_2	1.4	1.4	1.2	Ti	n.d
Cr_2O_3	0.1	0.1	n.d	Cr	n.d
MnO	n.d	n.d	n.d	Mn	n.d
FeO	2.7	2.4	2.1	Fe	0.7
CoO	n.d	n.d	0.1	Co	n.d
NiO	n.d	0.1	n.d	Ni	n.d
CuO	0.1	n.d	0.9	Cu	92.3
PbO	n.d	n.d	n.d	Pb	n.d
	96.6	94.9	87.6		94.6

Table 7 Slag Type I, bulk and prill analyses

Mineral texture
Rounded iron oxide dendrites in massive silicate (fayalite?).

Interpretation
Classic tap slag has few iron oxide dendrites, usually skeletal, with lath silicates in a glassy groundmass. This mineral texture indicates an iron-rich slag that cooled slowly. It is probable that it is a slag derived from puddling or a similar process.

Additional samples
Sample DWF88: Context 1006
This material was identified subsequently to the first slag investigation, as not fitting the established typology.

Morphology
Black, cindery friable slag, occurring as small lumps (100s gm maximum). It has some visual similarities to Slag Type B, but lacks the characteristic flat surfaces of that type.

Interpretation
Probably mineralised fuel, but may possibly be a carbonised cementing mix.

Sample DWF88: Context 931

Morphology
A randomly shaped lump (approximately 400 gm in weight) which was initially thought to be an iron ore.

Metallographic analysis showed that it was a fayalitic slag containing iron oxide dendrites in massive silicate with a glassy phase present at the silicate grain boundaries. It could therefore be a variant of Slag Type A, but with a higher iron oxide content and a slower cooling rate, or of Slag Type M. The morphology is more similar to Slag Type A.

Conclusions
Slag Types C, D, F (metal) and I were associated with the cementation process. Slag Type H was a pot-lid from the crucible process. Slag Types B and G were possibly associated with the cementation furnace. Slag Type E is a blast furnace slag. The genesis of Type A has not been ascertained.

Metallographic analysis of metallic samples
Dr Gerry McDonnell

Six metal artefacts were selected for metallographic analysis. Samples were removed using a jeweller's piercing saw, mounted in hot setting resin, and ground and polished in the usual manner (for details see McDonnell 1992). The mounted sections were examined in the as-polished condition and then etched in 2% Nital (nitric acid in alcohol). This programme was envisaged as a basic pilot study, restricted to metallographic analysis, to assess the research potential of the Derwentcote material. The majority of metal artefacts derive from post-abandonment contexts, with poor dating evidence for their fabrication.

The study aimed to examine and contrast structural iron with artefacts that were stock or products of the cementation process. In particular it was important to examine bar iron that was going to be carburised or had been treated in the furnace. There are only three modern published examinations of blister steel. Barraclough and Kerr (1973, also Barraclough 1984, 43 and plate 3) reported on some samples of the last blister steel produced in Sheffield in 1951. Samuels (1980, 187 and fig 53) discusses examples of blister steel, shear steel and crucible steel, but the samples are undated and unprovenanced. Rostoker and Dvorak (1988) discussed the cementation process and argued that, as well as producing blisters on the bars by evolution of gas from the slag inclusions, the process also removed most of this slag by conversion to metallic iron. They undertook some experimental work and included the analyses of a piece of steel from Jamaica, assumed to be of British origin, recovered during 'digging at the site of an old sugar mill' but for which no date was provided. They identified the steel as blister steel on the basis of the metallographic analysis and an analysis of a single slag inclusion. There are a number of reservations with this interpretation, and it may not be blister steel. If this were to be the case, there are no provenanced analyses of pre-1950 blister steel.

Results

The results are summarised in Table 8.

Artefact DWF 87/185
Flat bar. Length: 215 mm; width: 63 mm; thickness: 13 mm

The surface was corroded, with no obvious blisters. One end was chamfered, probably due to (hot?) cutting. The section was removed from the other end.

Examination in the unetched condition showed that the metal was clean and contained finely distributed small spheroidal slag inclusions. Etching showed that the sample had a varied microstructure; it was predominantly degraded coarse pearlite (μHV = 266, Plate 46), but with some proeutectoid ferrite (μHV = 128) and grain boundary cementite. The austenitic grain size indicated by the ferrite and cementite is large. The cementite in the pearlite was tending to coalesce and spherodise. The carbon content was hyper-eutectoid, but was hard to estimate given the coarseness of the microstructure. The examination indicates that this bar had been thoroughly carburised, but that prolonged overheating had caused the cementite to start to coalesce. All these features are in accordance with the cementation process. The coarse microstructure would have been removed during subsequent manufacture and treatments, and the steel could have been utilised for blades.

Artefact DWF 88/094

The bar is illustrated (Fig 25) and is stamped with a 'Hoop L' (Barraclough 1984a, 173). On the surface there was evidence of blisters.

The bar was clean with few slag inclusions. In the etched condition the microstructure was shown to be similar to 87/185, consisting of coarse pearlite (μHV = 260) and ferrite (μHV = 98) plus grain boundary cementite, with a carbon content > 1% C. In areas of corrosion on the surface the cementite still survived. The bar had been carburised, and it is interesting to note that the Swedish stamp was not removed prior to the bar being placed in the furnace, and that it survived the process.

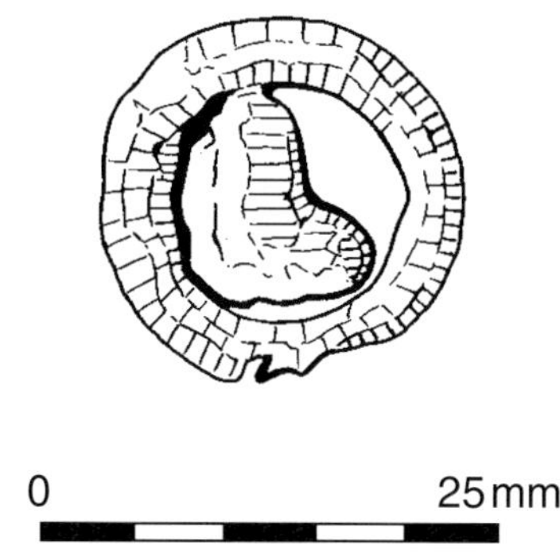

0   25 mm

Fig 25 'Hoop L' stamp

Artefact DWF 88/420

Flat, tapering bar. Length: 218 mm; width: 74 mm; thickness: 13 mm
The surface was lightly corroded, and there were no obvious blisters on the surface. Approximately halfway along the bar length, it had been drawn down to a few millimetres thickness. The thick end was chamfered, probably due to (hot?) cutting. The section was removed from the chamfered end.

In the unetched condition the metal was clean with some slag inclusions present. Etching showed the microstructure to comprise irregularly shaped ferrite grains (μHV = 112) varying in size from ASTM 45 – ASTM 1. In the ferrite grains along one edge a remnant phase (μHV = 123) was present which was probably due to segregation (of carbon?). Ferrite grain boundaries cut these phases and they had a semi-liquid/pseudo-dendritic appearance.

Artefact DWF 91/590

Section removed from lintel, south face of furnace (*Ch 5*).
Examination in the unetched condition showed the bar to contain numerous elongated slag stringers and a number of smaller slag inclusions. In the etched condition it showed a ferritic grain structure typical of wrought iron (Plate 48). A lack of obvious lamination in the *in situ* bar, compared to most other pieces of structural ironwork, had led to a field interpretation as steel, but this has been disproved by analysis; the bar was a piece of structural wrought iron.

Artefact DWF 91/787

Section from lintel, north face of furnace (*Ch 5*). In the unetched condition the metal contained no large slag stringers, but numerous slag inclusions were present. Etching revealed a mixed structure (Plate 49) ranging from small-grained to large-grained ferrite, and a carburised zone was present with a pearlitic microstructure and a maximum carbon content of 0.7% C. In the highest carbon areas austenitic grain boundaries are identified by formation of ferrite and widmanstatten ferrite.

The beam had been manufactured from a relatively clean iron, possibly of Swedish blister steel quality. The localised carbon rich area probably arose from localised carburisation rather than de-carburisation of a cemented steel bar.

As with 91/590, a lack of obvious lamination in the *in situ* bar, compared to most other pieces of structural ironwork, had led to a field interpretation as steel, but this has been disproved by analysis.

Sample	Type	Metallography	Interpretation
DWF 87/185	Flat bar	Degraded coarse pearlite	Product of furnace
88/094	Stamped bar	Coarse pearlite	Product of furnace
88/420	Tapered bar	Ferritic iron	Stock for furnace
91/590	Lintel	Ferritic iron	Structural iron
91/787	Lintel	Mixed	Structural iron

Table 8   Summary of metallographic analyses

Conclusions

Artefact 88/420 was the same size as the carburised bars. Its ferritic microstructure suggests that it was bar iron stock that was brought in for cementation but was rejected. Artefacts 87/185 and 88/094 were remnants of carburised blister steel bars, characterised by finely distributed small inclusions and a coarse pearlitic microstructure. Artefacts 91/590 and 91/787 were structural iron; they are proved to be wrought iron rather than the steel that had been suspected on site.

These initial results demonstrate that the Derwentcote metal samples are a valuable research archive. Further analysis would greatly benefit the study of the cementation process. In particular, detailed characterisation of the variation in composition of inclusions may enable blister steel to be recognised more readily in contemporary artefacts.

The animal bone

The collection of mammal and bird bone was small, and over half was deposited after closure of the furnace (reflecting the concentration of the excavation strategy on post-abandonment deposits). The few identifiable bones from the construction period were all sheep metapodials. The assemblage from the industrial phase was also dominated by sheep metapodials, all in poor condition, with no domestic food waste. In contrast, the assemblage from post-abandonment infills (largely in the ashpit) was dominated by food waste, the bones indicating good-quality cuts of butchered mutton, beef and poultry (mainly domestic fowl, but including two duck bones from a large breed). Some fragments were heat-parched. The recent assemblage was also of butchered food debris, of cattle, sheep and poultry; some were heat-parched or burnt.

An anomalous small group came from a relatively late (but poorly-stratified) posthole in the north-east exterior. This consisted of fowl bones (probably all from one bird), together with the only pig bones from the site, from a very young animal. This could be food debris from a special meal of chicken and sucking pig, though the interment of a still-born piglet is also possible.

The high proportion of sheep metapodials in the construction and industrial phases suggests a non-food function, possibly for pegging tiles or slates (Ryder 1983, 717–720), though pantiles were normally hung rather than pegged. In contrast, the post-industrial assemblage was dominated by good-quality roasting meat joints. This may have originated from a closing-down meal, or have been redeposited domestic rubbish (from Derwentcote House?).

A detailed report is filed in the Site Archive (Gidney 1991).

8

DISCUSSION AND SYNTHESIS

The historical evidence indicates that Derwentcote Forge was built in 1718–19 by Ralph Reed, passed to the Thomlinsons in 1721, and from them to Cuthbert Smith and his partners in 1733. The furnace is first mentioned in 1742, and was probably built by Smith and his partners, in or after 1733. These partnerships are not well known, and their place in the regional and national iron and steel industries is unclear.

The antiquity of the links between Derwentcote and Blackhall Mill are uncertain, since the ownership of Blackhall Mill during this period is unknown. Photographs of Blackhall Mill (notably Beamish Museum negative 15397) indicate that its structure was very similar to Derwentcote, so much so as to suggest a common builder; the site was founded in 1687, but the furnace need not have been original, and Angerstein's description in 1753 (Barraclough pers comm) could be read as implying that it was a relatively recent addition or rebuilding. From at least the time of Blenkinsop's attempted eviction (the late 1740s?), the workforces of Blackhall Mill and Derwentcote were closely linked, regardless of the changing linkages of the owning partnerships.

The combination of historical evidence with the firebrick stamps indicate that the furnace was in use until at least 1875; while the site as a whole remained in use until 1891, the documentation suggests that the emphasis was on crucible steel production, and it is uncertain whether or not the cementation furnace remained in use.

The iron forge was built in 1718–1719, and was accompanied by a corn mill (probably that recorded from 1660 onwards) until between 1742 and 1748. In the 1780s it was at the forefront of the developing technology of stamping and puddling, but this picture of energy was not maintained, and it closed in the 1850s. However the forge site remained in use for steel forging, and for crucible steel production, until 1891.

Turning to the site evidence, it may be noted that the (limited) investigation of construction levels produced no evidence for any structures or stratigraphic accumulation preceding the construction of the standing furnace, which was almost certainly the first structure on the site. However the presence of fragments of burnt stone and blast furnace slag in the fossil soil cut by the construction deposits does indicate some earlier activity nearby (*see below*). Both structural and stratigraphic evidence indicate that the furnace was the earliest of the surviving structures, though the presence of tusking stones on its corners, and the lack of any stratigraphic build-up between the foundation trenches of the furnace and of the ancillary buildings, suggest that the interval was short.

The structure of the furnace itself can be divided into an outer stone casing, and the internal linings and structures of the firegrate, chest compartment, and the floor of the cone interior. The outer casing can be accepted as being original work from the earlier eighteenth century, with only limited amounts of later repairs and alterations. As such, it forms by far the earliest known surviving cementation furnace in Britain, if not in the world.

As already described (*Ch 5*) the lower parts of the outer structure displayed some odd features, both in construction and in condition. The keying of the buttresses to the main structure was inconsistent (and is interpreted as block-keying), the structure was disturbed by wide sub-vertical cracks and in places by overhangs interpreted as horizontal shearing, and the faces of the stone and mortar were often obscured by encrustations and weathering. All these features were concentrated on the lower rectangular part of the structure; they tended to die out downwards near ground level, and only the sub-vertical cracking extended onto the upper cone of the furnace. This area, outside the chest compartment, is precisely that most exposed to heat flow from the interior of the furnace, and it is likely that the block-keying was intended to allow for the thermal movement of the structure, while the cracking and shearing resulted from thermal movement not fully allowed for in the design. The weathering and encrustation may also have been the result of prolonged heating, augmented perhaps by percolation of sulphates and other chemicals from the coal fumes circulating in the interior.

The beams along the face of the furnace, built into the tops of the buttresses of the west face, and the sockets in similar locations on the east face, are unusual features; the sawn-in rebates imply that they may have been jointed to similar beams along the north and south faces (where they would have rested on the tops of the buttresses). There are no parallels known (from either contemporary or recent sources) on cementation furnaces. However a parallel can be found in the timber strapping known from sixteenth and seventeenth-century blast furnaces (Cleere and Crossley 1985, 188, 244; Crossley 1990, 158), and it is suggested that the Derwentcote features formed the remains of a similar strapping round the 'shoulder' of the furnace, to resist the expansion caused by internal heating. This strapping, if such it was, was largely removed at some date; the distribution of heat-reddening on the east and west faces suggests that the beam may have been ignited by the heat from the furnace, possibly penetrating through a crack in the structure. However it is not possible to ascertain how far the cracking of the furnace preceded (and may have caused) the removal of the strapping, and how far it post-dated (and resulted from) the failure of the strapping.

The top of the furnace cone had been broken off throughout its circumference, leaving no direct evidence of its original height or circumference. Comparison with Blackhall Mill (Plate 3)(as shown on Beamish Museum Negative 15397) suggests an original height of c 9 m from ground level (1 m more than survives), and a top diameter of slightly less than 1 m.

In contrast with the outer casing, the internal linings and structures of the furnace are unlikely to be original (with the exception of the stone walls of the ashpit and the cone interior, which are integral to the outer casing). The brick vault and (presumably) side-walls of the cementation chamber were separated from the outer casing by a narrow gap, and could be removed and replaced without damage to the outer structure; the cementation chests and the flues round them could clearly be assembled and dismantled *in situ*, without structural damage to the chamber that contained them. Barraclough (1984a, various references) confirms that cementation furnace chests and linings had a limited working life due to the severe conditions to which they were exposed, and were regularly replaced. The firebrick parts of the ashpit walls also appear to be a replacement, though in this case perhaps not allowed for in the design.

The surviving firegrate, chest compartment, chests, and cone floor should not therefore be considered as original early eighteenth century features, but as later replacements fitted into the eighteenth century casing. Given that they themselves show evidence of repair and alteration, it is likely that the outline of the present internal structure had been in use for some considerable time before final abandonment, with considerable patching and some minor alterations (notably the enlargement of some loading holes, and the insertion of the lower bar hole in the north end). It may therefore be suggested that the surviving internal linings date from the (postulated) re-opening of the furnace in the 1830s, with later alterations. The chests are likely to have had a shorter working life; Andersson (translated in Barraclough 1984a, 41–2) quotes a working life of $1\frac{1}{2}$–2 years for the chests of a Newcastle furnace in 1767, and the large amounts of discarded and vitrified refractory 'chest' sandstone found in excavation (and scattered over the dumps to the north of the furnace) confirm that the Derwentcote chests were replaced at quite frequent intervals. The insertion of the single lower bar hole (seemingly cut into the western chest) may therefore have dated from only just before the closure of the furnace.

Earlier linings and internal structures may of course have differed from those that survive, both in form (within the constraints imposed by the outer casing) and in material. In particular, it is doubtful whether firebrick would have been used for the chest compartment lining in the early eighteenth century (the early history of firebrick has not been adequately studied, but the author's impression is that it was not a normal lining material until late in the century); the occasional finds of glazed red sandstone in the excavation, and in repairs and alterations to the buildings, may be remnants of the original lining material.

The detailed forms and functions of some features of the furnace merit discussion. The ashpit presumably had the capacity to hold all the ash produced during a firing of the furnace; during a firing it would probably have been completely covered over by the iron plates at floor level (since otherwise there would have been no access to the ends of the firegrate for stoking), the ash being cleared out between firings. This implies that there would not have been a free flow of air from the Northern and Southern Buildings into the ashpit, to supply the fierce draught needed by the furnace. It is therefore likely that the culvert running north from the base of the ashpit functioned at least in part as an air conduit, rather than as a drain. The use of a culvert from the outside air to carry the draught to a furnace has parallels in the early eighteenth century Bristol brass industry (Day 1988, 25). The opening of the culvert was an original built-in feature, but its excavated external continuation was not original. This was presumably due to rebuilding, for which there was some evidence in excavation.

The iron bars across the top of the ashpit were too widely spaced to be the bars of a grate, and lay appreciably below the base of the surviving mass of vitrified ash in the firegrate area and of the intense vitrification of the sidewalls; they were presumably therefore the supports for longitudinal removeable firebars. Removal of these bars probably provided the normal access to the interior of the chest compartment (as described for a local furnace by Andersson in 1767 (Barraclough 1984a, 40)), since there was no separate access doorway, and the original loading holes were probably all too small for man-access; the northern loading holes as later enlarged were just big enough for access, though removal of the firebars might still have been the easier route. The ends of the firegrate were presumably closed by firedoors; the large iron frame recovered from the surface of the northern ashpit may have been the doorframe for one of these.

The chest flues provided for a free circulation of flame around the cementation chests; this flame and that rising directly between the chests was reverberated back onto the tops of the chests by the vault of the compartment, before being carried away by the cone flues. As in contemporary descriptions of Tyneside furnaces, the tops of the chests were sealed by a layer of sand, producing the masses of fritted sand (Type C 'slag') which form a major component of the process waste assemblage. The small holes ('bar-holes') just above the ends of each chest were presumably used for passing bars of iron into, and blister steel out of, the furnace; the blocking of the northern bar-holes may reflect a design error in that the distance from their openings to the north wall of the Northern Building was less than the length of the Northern Building (and therefore inadequate for loading of a full-length bar), though there is some evidence that they were used. For similar reasons of access, the loading holes in the sides of the furnace could not have been used for loading of bars; the stratigraphic accumulations outside the western holes show that they were used *inter alia* for loading of sand, and perhaps also of charcoal.

The small internal chimneys by which the cone flues vented into the cone interior were presumably to promote a strong draught, and avoid blockage of the flues by any ash, soot, or debris accumulating on the cone floor. The inclined shafts from the tops of the firegrate arches to the cone floor may have been merely to clear fumes from inside the Northern and Southern Buildings; if they contained any

arrangement for controllable blocking (for which no evidence survived) they could also have been used to regulate the draught through the furnace. The main doorway into the cone was presumably for access for maintenance and repair; it was probably blocked during firing by an iron-plated door, perhaps that recovered from the area to its east. The small blocked opening into the cone, in the south-east octant, is harder to explain, though it was closely paralleled at Blackhall Mill (Beamish Negative 15397); it may possibly have carried the rope for a pulley to raise heavy blocks inside the furnace, during construction and major repairs.

The three standing ancillary buildings were very similar in construction to each other, but not to the furnace. However the presence of tusking stones indicates that the Northern and Southern Buildings were part of the original design, the difference in construction presumably being functional rather than chronological. The Eastern Building was structurally secondary to the Southern, but the presence of paired doorways in the east wall of the Southern Building (one to the Eastern Building and one to the exterior) implies that the Eastern Building was also part of the original design. The rather deep construction terrace for the northern part of the Southern Building may imply that a shorter building with a lower floor level was originally envisaged; if so, the design was rapidly changed. The northern buttress was also very similar in construction to the ancillary buildings, and was constructed before the landscaping of the area; it was almost certainly an original feature, despite its abutment to the Northern Building.

The junctions of the Northern and Southern Buildings with the furnace were rather odd. Three of the four junctions clasped the corner of the furnace; this may have been to allow a slightly greater internal width, or to protect the corners of the timber strapping within the thickness of the walls (at the expense of access for repair). The fourth junction (of the west wall of the Southern Building) was against a buttress of the furnace, missing the corner completely; the absence of tusks implies that this was deliberate rather than being a setting-out error, but no explanation can be offered.

An unusual feature of the construction was the use of fine sand rather than mortar as a bedding medium; this was consistent throughout the primary masonry of the ancillary buildings. It is likely that the buildings were consistently pointed externally with sandy Type V mortar, though survival of this was patchy. Similar sand bedding was used in one phase of Derwentcote Farm, and it has also been noted by the author in a wall (of unknown date) at Winlaton Mill ironworks. It may therefore be a local vernacular feature. The sandstone used in the ancillary buildings was softer and browner than that used for the furnace iteif, and the masonry was smaller and less regular. It is likely that some of this stone was obtained from Derwentcote Quarry, the better-quality stone perhaps being brought from a greater distance.

The buildings had clearly been re-roofed, perhaps several times, though the positions of at least some trusses in the Southern Building had remained constant. The final roofs were entirely of pantile (including a few glass pantiles). The presence of pantile fragments in early stratigraphic contexts (including the pointing of the Northern Buttress), coupled with the rarity of other roofing materials in the finds, strongly implies that the roofs had always been pantiled.

The only other architectural feature to note is the differing treatment of the northern and southern gables; the southern gable finished in a plain close verge, whereas the northern appears to have been carried up into a low half-parapet, decorated externally with a border of ashlar blocks (a local vernacular trait; see for example Emery *et al* 1990, 141, 143). The difference is likely to have been symbolic rather than functional, presumably to provide a more impressive facade on what is otherwise a very functional group of buildings.

The Northern Building showed no signs of alteration, and was probably used for storage and as a stoking area. The beam sockets in its walls may have held a loft, and perhaps some other timber structure. The Eastern Building may well have been an office, with its relatively good light and fireplace.

The Southern Building had a more complex history, in that its south half had been considerably altered; the east wall had been rebuilt from footing level, and the window in the west wall had been moved. There is evidence to suggest that this rebuilding was the result of a fire, and also that as originally built it had a third doorway in the east wall. The alterations cannot be closely dated, but probably occurred within the working life of the furnace. The large opening in the south wall is as likely to have been a loading opening as a window, its base being close to external ground level before the construction of the pathway trench. Its blocking probably dated from after the closure of the furnace (from the stratigraphy of the probable builders' debris), as did that of the western doorway.

Much of the stratigraphy excavated inside the Southern Building could not be interpreted, except insofar as that it represented a sequence of thin compacted earth accumulations, with traces of several stone floors. The slag deposits in the southeast quadrant suggest smithing or the storage of iron/steel, and the tree-trunk embedded in the floor may well have formed the base for an anvil (Crossley 1990, 167); complex pit 1138 to its west may have held the base for a second anvil, or possibly a hammer. The iron floor plates may also have been intended to resist damage from hot metal, though they may represent merely the re-use of scrap metal. It is therefore likely that the southern part of the building was used for minor forging or smithing, though it should be noted that the process was definitely not water-powered. More specifically, the tree-stump may have been used as an anvil base into which a chisel could be embedded for cutting the iron bars to length for cementation, as described by Rees in 1819 (Cossons (ed) 1972, 251; also quoted by Barraclough (1984a 189)); it lay 3.5–3.8 m from the south wall of the building, as compared to the 3.6 m length of the chests.

The main feature of note in the north end of the building was the pair of semi-circular iron plates. These were recovered from a post-abandonment context, but there was evidence to suggest that one (at least) had been set in a very similar position during use; the brick alteration to the top step of the ashpit may have been to accommodate the plates, set side-by-side to form a full circle. Several functions have been considered for these plates; given their location, and the annular wear on one face, the preferred interpretation is that they were the base for a charcoal-grinding mill, formed by an edge-runner stone running on the paired plates, and powered by human effort (or possibly by horse, though no evidence for a gin-circle was recovered). The use of a charcoal-grinding mill is mentioned by Angerstein, though the type is not specified (Barraclough 1984a, 41). This interpretation is given some support by the amount of charcoal dust present in the northwest quadrant of the building. It is possible that the upper parts of the mill could be dismantled when not in use, to allow unimpeded access for stoking the furnace during a firing.

The identifiable agricultural alterations to the building can be dated to *c* 1900 or later on finds evidence (*Ch 7*); the building may therefore have been abandoned, or re-used without major alteration, for some thirty years before this conversion. It may be noted that much of the infilling of the ashpit did precede the alterations (on stratigraphic evidence). The suggestion, from finds evidence, that this infill contains the remains of a closing-down party or 'wake' is intriguing, though unfortunately more prosaic explanations are possible.

The pathway trench round the south end of the building was stratigraphically secondary to the construction terrace, and ceramic evidence indicates that it was probably of nineteenth century date. It probably served to reduce the flow of moisture into the building, more than for access, and would have obstructed access to the opening in the south wall (which may have ceased to be used for loading, remaining open as a window). The construction of the South-east Building slightly post-dated that of the pathway trench (on stratigraphic evidence). This structure was shown as an enclosure rather than a roofed building on the 1st edition OS plan (dating it and the pathway trench to before 1856), and was probably an external yard.

In contrast, the two timber-framed buildings (945 and 1020) at the north end of the complex were not shown on the 1856 OS plan; given their reasonably early stratigraphic level, they are likely to have been abandoned before this date. Deposits of sand in Building 945 may point to interpretation as a sand store, but were too small to give strong evidence. Similar arguments, for both dating and function, apply to the timber structure west of the Northern Building. The kerb to the east of the furnace may have been the remnant of a further structure, in this case probably a flimsy timber lean-to sheltering the northern loading hole.

The finds assemblage was in many ways dis-appointing as a source of information, no doubt because the interior and immediate surroundings were kept clean during use; excavation of the tips on the hillslope to the northeast might yield very different results. The most useful dating evidence (for the closure of the furnace) proved to come from the glass and firebrick stamp assemblages. In the absence of any modern study of the development of firebrick, the significance of the visible fabric variations cannot be assessed; the assemblage would be a useful source for future investigation of this topic.

The assemblage of process residues forms a valuable type-series for the archaeometallurgy of steelmaking. To judge by this assemblage, cementation furnace assemblages should be readily identifiable by the combination of refractory sandstone (often vitrified), and fritted sand masses (Slag Type C), though doubtless the detailed form of both materials will be affected by local variations (for example the use of wheelswarf for the chest-sealing material in the Sheffield area produced a material ('crozzle') distinctly different from the Derwentcote fritted sand (Barraclough 1984a, 42–3). The presence of crucible sherds (Slag Type H), and particularly their distinctive lids, should similarly be diagnostic of crucible steelmaking. Vitrified brick (Slag Type D) and vitrified ash/hearth lining residues (Slag Type B) will presumably also be normal components of cementation steelmaking assemblages, but are unlikely to be diagnostic, since other coal-fuelled processes will presumably produce similar materials.

The presence of Slag Type A is a major interpretative problem, since its process origin has not been established, and it is a major component of the assemblage. Its occurrence in very large quantities in late and post-abandonment contexts, yet with appreciable amounts at much earlier stratigraphic levels, suggests that it may have multiple process origins. Local observation indicates that visually-similar material is common as road metalling in the Tyneside/Derwentside area. It is therefore suggested that the bulk of the type derives from a late puddling/or steelmaking process (not necessarily at Derwentcote, given the possibility of introduction as road metal), but that the type as currently defined includes earlier material from the finery and/or puddling process, as practiced at Derwentcote Forge.

In contrast, Type E was present in small quantities, and in early contexts, and can be reliably identified (as blast furnace tap slag). However the presence of this material at Derwentcote is unexpected. It may have been introduced as road metalling (though none was found *in situ*), perhaps in conjunction with pig iron brought in to the forge, but it also possible that it originated from a blast furnace nearby; the forge area would be topographically suitable. The evidence is too weak to establish this as a probability, but the slight possibility should be borne in mind in any future work. It may be noted in this context that the source of pig iron for the forge in its earliest years is not known; it is not clear whether Allensford furnace remained in blast late enough to have supplied Derwentcote.

Glossary

Bar iron

Wrought iron (from the bloomery or finery) formed into plain bars, suitable for trading and transport. Bar iron formed the raw material both for production of wrought-iron artefacts (by smithing or forging) and for cementation steelmaking.

Bloomery

An iron smelter producing metallic iron in the solid state direct from the ore, rather than producing liquid cast iron. The smelting furnace produced a 'bloom' of iron mixed with slag, which was then smithed to expel the slag and form bar iron. The process was used from the Iron Age to final abandonment in the early eighteenth century, with a range of technical modifications during this period. In particular, the process was mechanised during the Middle Ages by the use of water power for operating the bellows and/or hammers. Although the normal product was low-carbon wrought iron, the smelting process could sometimes be modified to incorporate some carbon into the metal, resulting in the production of steel.

Blast furnace

A tall iron-smelting furnace for producing cast iron. Ore and fuel were charged into the top of the furnace, and a powerful air blast (normally from water-powered bellows) was blown in at the base. Due to the high temperature and prolonged contact between iron and fuel, the iron absorbed sufficient to form the high-carbon alloy cast iron, which was molten at operating temperature and could therefore be run out of the furnace allowing continuous operation.

Blister steel

The immediate product of the cementation furnace. The form of the wrought iron bars which had been converted remained, but the surface was marked by blisters (due to reaction of the diffusing carbon with streaks of slag in the bars). The carbon content was highest near the surface, the mean content varying from 0.6% for spring steel to 1.35% for the hardest forgeable steels, and 1.6% for steel intended for crucible melting. Blister steel was brittle due to its crystal structure, and needed forging before it could be used. For the lower-carbon steels (roughly 0.6 – 0.9% mean carbon content) this was done by simple hammering or (later) rolling, followed by working into springs or cheap cutlery. Higher- carbon steels were forged into shear steel, or melted to form crucible steel.

Case hardening

A form of small-scale cementation, in which a pre-formed wrought-iron artefact was heated in a deep charcoal bed, sufficient diffusion of carbon occurring to convert the surface of the artefact to hard steel, over a softer but more resilient core (Barraclough 1984a, 24).

Cast iron

An alloy of iron containing 3 – 4% carbon, produced by the blast furnace. Cast iron is hard but brittle, and cannot be forged; it can be used directly for preparing castings (its melting point of 1130 – 1300° C was relatively easy to achieve), or decarburised in the finery to produce steel or wrought iron.

Cementation

The diffusion of one solid into another, without liquefaction; in practice this phenomenon only occurs at detectable rates at high temperatures. In steelmaking, the diffusion is of carbon (from charcoal) into bar iron, thus converting the iron to blister steel. The process had to be conducted in the absence of air (since otherwise both the carbon and the iron would oxidise), and took about a week at temperatures of 1050 – 1100° C, this being the highest temperature that could be achieved without risk of melting the most highly- cemented parts of the steel (1.7% carbon steel melts at 1130° C).

Crozzle

The fired sealing material from Sheffield cementation furnaces. In Sheffield, chests were sealed with the sludge from cutlery grinders' troughs, consisting of a mixture of sandstone and steel particles. This fired to an impervious refractory cement, which formed distinctive jagged lumps (sufficiently sharp to be used as intruder-proofing on walls) when broken up to retrieve the steel (Barraclough 1984a, 42). So far as is known plain sand was normally used in the Newcastle area, and the finds assemblage from Derwentcote contains fritted sand rather than crozzle.

Crucible steel

Until the mid eighteenth century, it was not possible to melt blister steel to produce castings, due to the lack of furnaces and refractory materials capable of withstanding the temperatures and chemical conditions needed; 1.7% carbon steel starts to melt at 1130° C but is not fully liquid until over 1400°C, and a 0.6% carbon spring steel starts to melt at 1350° C and is fully molten at 1500° C. The crucible process was developed by Benjamin Huntsman in Sheffield around 1740. It used a small coke-fuelled furnace set in the floor of a workshop, with a strong air draught from an ashpit at cellar level, and a pre-fired crucible of special refractory clay to contain the steel for melting.

Until the mid-nineteenth century, crucible steel was made entirely by melting blister steel already cemented to the right   carbon content. After 1855, however, phosphorus-free Swedish cast iron became available, and it became possible to use puddled steel or mixtures of cast and wrought iron in the crucible, thereby avoiding the high cost of cementation.

Feesing house   The building attached to a cementation furnace, and used for storage and as a working area. The term is known from Tyneside eighteenth century documents, and may have been a local vernacular.

Finery   The hearth in which cast iron was converted into wrought iron by burning-out the carbon content. A low open hearth was used, with charcoal fuel and an air blast from water-powered bellows. Pigs of cast iron were fed in and melted, so that they dribbled down through the oxidising air blast, and solidified into a lump of carbon-free wrought iron, which was then hammered and reheated repeatedly (normally in a second hearth within the same building) to form it into bar iron. The works on which this occurred was known as a finery forge.

Luting   A refractory bedding or sealing material, typically fireclay, used in the hotter parts of a hearth or furnace where mortars would melt or react with their surroundings.

Öregrund iron   Bar iron from the Öregrund or Dannemara area of Sweden. This iron was free from phosphorus and sulphur, and was very carefully forged, and was considered to be the essential starting point for cementation steelmaking.

Puddling   The process of burning-out the carbon from cast iron in a coal-fuelled reverberatory furnace (very different in design from the cementation furnace, despite the shared reverberatory principle). The process was developed in the late eighteenth century, and replaced the finery as the normal method of producing wrought iron. In the nineteenth century, the process was also used with increasing success for producing steel direct from cast iron, by burning out only part of the carbon.

Refractory   A material that is resistant to melting and chemical attack at high temperatures, and is therefore suitable for crucibles and furnace linings. Normally either a high-silica sandstone or a fireclay.

Reverberatory furnace   A furnace in which the fuel was separated from the material to be heated, the heat being carried by flames which were reflected ('reverberated') down onto the material by a vault. Reverberatory furnaces could use any long-flame fuel (in practice coal in Britain), and relied on a strong air draught, rather than bellows), to draw the flames through the furnace.

Shear steel   Very finely-laminated steel, formed by repeatedly forging down bundles of blister steel bars, cutting up and reforging. This largely homogenised the carbon content (typically 0.9 – 1.15%), while retaining a laminar structure. Shear steel is believed to have been introduced into Britain at Blackhall Mill, and was a speciality of the Newcastle area. In the eighteenth and nineteenth centuries it was also referred to as 'German steel', though in earlier centuries this term had a different meaning.

Wrought iron   Commercially-pure iron, with a negligible carbon content but with a laminar structure containing streaks of slag. Wrought iron can be forged and welded, but cannot be cast. It could be produced directly from the bloomery, or indirectly in the finery from cast iron. It was the main form in which iron was used until the mid-nineteenth century, since when it has been replaced for most purposes by mild steel.

Bibliography

Ashurst, D, 1987 Excavations at the 17th – 18th century glasshouse at Bolsterstone and the 18th century Bolsterstone Pothouse, Stocksbridge, Yorkshire, *Post-Medieval Archaeol*, **21**, 147 – 226

Awty, B G, 1957 Charcoal ironmasters of Cheshire and Lancashire, 1600–1785, *Trans Hist Soc Lancashire Cheshire*, **109**, 71 – 124

Bailey, J, 1810 *General view of the agriculture of the county of Durham*

Baker, J C, 1984 *Sunderland pottery*, Newcastle upon Tyne

Barraclough, K C, 1984a *Blister steel — the birth of an industry*, London

Barraclough, K C, 1984b *Crucible steel — the growth of technology*, London

Barraclough, K C, and Awty, B G, 1987 Denis Hayford: An early steel master, *Hist Metall*, **21**, 1, 16 – 17

Barraclough, K C, and Kerr, J, 1973 Metallographic examination of some archive samples of Steel, *Journ Iron Steel Inst*, **211**, 470 – 474

Bell, R C, 1971 *Tyneside pottery*

Bennett, G, Clavering E, and Rounding, A, 1990 *A fighting trade: rail transport in Tyne coal 1600 – 1800*, Gateshead

Billington, W, 1825 *A series of facts, hints, observations, and experiments on the different modes of raising young plantations of Oak…* London

Bolton, E, 1990 *The clay tobacco pipes from excavations at Derwentcote furnace, 1987 – 88* (typescript report filed in Site Archive)

Bown, L, and Nolan, J, 1990 The Finds, in the Castle of Newcastle upon Tyne after *c* 1600, (by J Nolan), *Archaeol Aeliana*, 5 ser, **XVIII**, 79–126 Nolan 1990, 107 – 118

Buckley, F, 1927 Potteries on the Tyne, and other northern potteries during the eighteenth century, *Archaeol Aeliana*, 4 ser, **IV**, 68 – 82

Chard, M, 1990 *Derwentcote: the glass* (typescript report filed in Site Archive)

Cleere, H, and Crossley, D, 1985 *The iron industry of the Weald*, Leicester

Cossons, N (ed), 1972 *Rees's manufacturing industry (1819 – 20)*, Newton Abbot

Cranstone, D, 1989 *Derwentcote forge and furnace: the history* (certificate thesis, Uni Newcastle)

Cranstone, D, 1991 Winlaton Mill Ironworks, *Archaeol North*, **2**, 22 – 24

Crossley, D, 1990 *Post-medieval archaeology in Britain*, Leicester

Davison, P J, 1986 *Brickworks of the North East*, Gateshead

Dawson, G, 1971 Pottery classification, in Norfolk House, Lambeth: excavations at a Delftware kiln site, 1968 (by B J Bloice), *Post-Medieval Archaeol* **5**, 99 – 159

Day, J M, 1988 The Bristol brass industry: furnace structures and their associated remains, *Hist Metall*, **22** No 1, 24 – 41

Emery, N, Warner, J, and Pearson, A, 1990 Causeway House, Northumberland, *Archaeol Aeliana*, 5 ser, **XVIII**, 131 – 149

Geological Survey of Great Britain, 1982 *Geological Map, 1:10,000, Sheet NZ 15 NW*

Fell, Alfred, 1908 *The early iron industry of Furness and District* (Ulverston, reprinted Frank Cass and Co, 1968)

Flinn, M W, 1955 Industry and technology in the Derwent Valley of Durham and Northumberland in the eighteenth century, *Trans Newcomen Soc*, **29**, 255 – 262

Flinn, M W, 1962 *Men of iron*, Edinburgh

Gidney, L J, 1991 *The animal bones from excavations at Derwentcote furnace* (typescript report filed in Site Archive)

Gurcke, K, 1987 British firebricks in North America, *Hist Metall*, **21**, 1, 18 – 24

Harrison, J K, 1969 The Derwentcote steel furnace, *Bull Ind Soc North East*, **9**, 12 – 18

Hulme, E Wyndham, 1929 Statistical history of the iron trade of England and Wales, 1717–1750, *Trans Newcomen Soc*, **9**, 12 – 35

Jarrett, M G, 1964 Makers of clay pipes recorded in North-Eastern England, II, in *Archaeol Aeliana*, 4 Ser, **XLII**, 255 – 260

Jenkins, R, 1935 The Hollow Sword Blade Company and sword making at Shotley Bridge, *Trans Newcomen Soc*, **15**, 185 – 194

Jennings, S, 1981 *Eighteen centuries of pottery from Norwich*, E Anglian Archaeol, **13**, Gressinghall

Killick, D J, and Gordon R B, 1987 Microstructures of puddling slags from Fontley, Roxbury, Connecticut, USA, *Hist Metall*, **21**, 1, 28 – 36

Lewis, M J T, 1970 *Early wooden railways*, London

Linsley, S M, and Hetherington, R, 1978 A seventeenth century blast furnace at Allensford, Northumberland, *Hist Metall*, **12**, 1, 1–11

Linsley, S M, 1992 The road past Killhope and some other orefield turnpikes, in *Men, mines and minerals of the North Pennines*, Killhope (ed Chambers)

McDonnell, G, 1992 The metallurgical analysis of the ironworking slags and iron artefacts, in *Anglo-Scandinavian Ironwork from Coppergate, York*, (Fascicle 17/6), (by P Ottaway)

Meigh, E, 1972 *The story of the glass bottle*, Stoke on Trent

Mills, D A C, 1982 *Geological notes and local details for 1:10,000 sheets NZ 15 NW, NE, SW, and SE (Chopwell, Rowlands Gill, Consett, and Stanley)*, Keyworth

Mott, R A, 1983 *Henry Cort: the great finer*, (ed P Singer), London

Parsons, J E, 1964 The archaeology of the clay tobacco-pipe in North-East England, *Archaeol Aeliana*, 4 ser, **XLII**, 231–254

Percy, J, 1864 Metallurgy, **II**, 3, *Iron and steel*, London

RCHME, 1990 Archive plans and reports National Archaeological Record No NZ 15 NW 24

Riden, P, 1993 *A gazetteer of charcoal-fired blast furnaces in Great Britain in use since 1660*, Cardiff, 2 edn

Ross, C, 1982 *The development of the glass industry on the rivers Tyne and Wear, 1700–1900*, unpubl PhD thesis, Uni Newcastle

Rostoker, W, and Dvorak, J, 1988 Blister steel = clean steel, *Archaeomaterials*, **2**, 175–186

Ryder, M L, 1983 *Sheep and man*, London

Samuels, L E, 1980 *Optical microscopy of carbon steels*, American Soc Metals

Spencer, T, 1864 On the manufacture of steel in the northern district, in *Industrial resources of the Tyne, Wear, and Tees*, by W G Armstrong, I Lothian Bell, J Taylor, and Dr Richardson, 120–125

Surtees, R, 1820 *The history and antiquities of the County Palatinate of Durham*, **II**

Tylecote, R F, 1986 *The prehistory of metallurgy in the British Isles*, London

Plate 3   Blackhall Mill steel furnace in the early twentieth century
(©Beamish North of England Open Air Museum. Neg no. 15397)

Plate 4   Derwentcote Forge in the 1860s or 1870s, from the west. The figures are probably the manager and his
family, and the rank of chimney stacks in the top left background are probably those of the crucible melting furnaces
(©Beamish North of England Open Air Museum. Neg no. 12663)

*Plate 5   The furnace in the 1960s, showing the remains of the roof
(©Beamish North of England Open Air Museum. Neg no. 7655)*

Plate 6   The furnace and buildings as cleared for recording, from the south-west

Plate 7   Furnace west face: loading holes (open and blocked) and central buttress

Plate 8   Furnace west face: remains of timber strapping beam at north-west corner

Plate 9   Furnace west face: remains of timber strapping beam at north-west corner

Plate 10   Furnace east face after excavation of east centre area

Plate 11   Furnace east face: southern loading hole

Plate 12   Furnace east face: blocked strapping beam socket in southern buttress

117

Plate 13   Furnace south face: firegrate arch and bar-holes

Plate 14   Furnace north face: inserted bar-hole on west side of firegrate arch

Plate 15   Furnace: the cone from the south in 1987

Plate 16   Furnace: ashpit, southern steps from the north

Plate 17   Furnace: ashpit, southern steps from the west

Plate 18   Furnace: interior of chest compartment from the north (J Stanger)

Plate 19   Furnace: chest compartment, opening of cone flue (J Stanger)

Plate 20   Furnace: cone interior, northern end from above

Plate 21   Furnace: cone interior, south end from north (2 metre scale)

Plate 22   Furnace: cone interior, inner face of opening

Plate 23   Northern Building: west wall, exterior face (2 metre scale)

Plate 24   Southern Building: west wall exterior, altered window

Plate 25   Northern Building: interior as excavated, from the south-east

Plate 26   Southern Building: interior as excavated, from the south. The wall-tops have been given temporary winter protection by English Heritage

Plate 27   Southern Building: interior during excavation, from the north

Plate 28   Southern Building: tree-stump in primary make-up

125

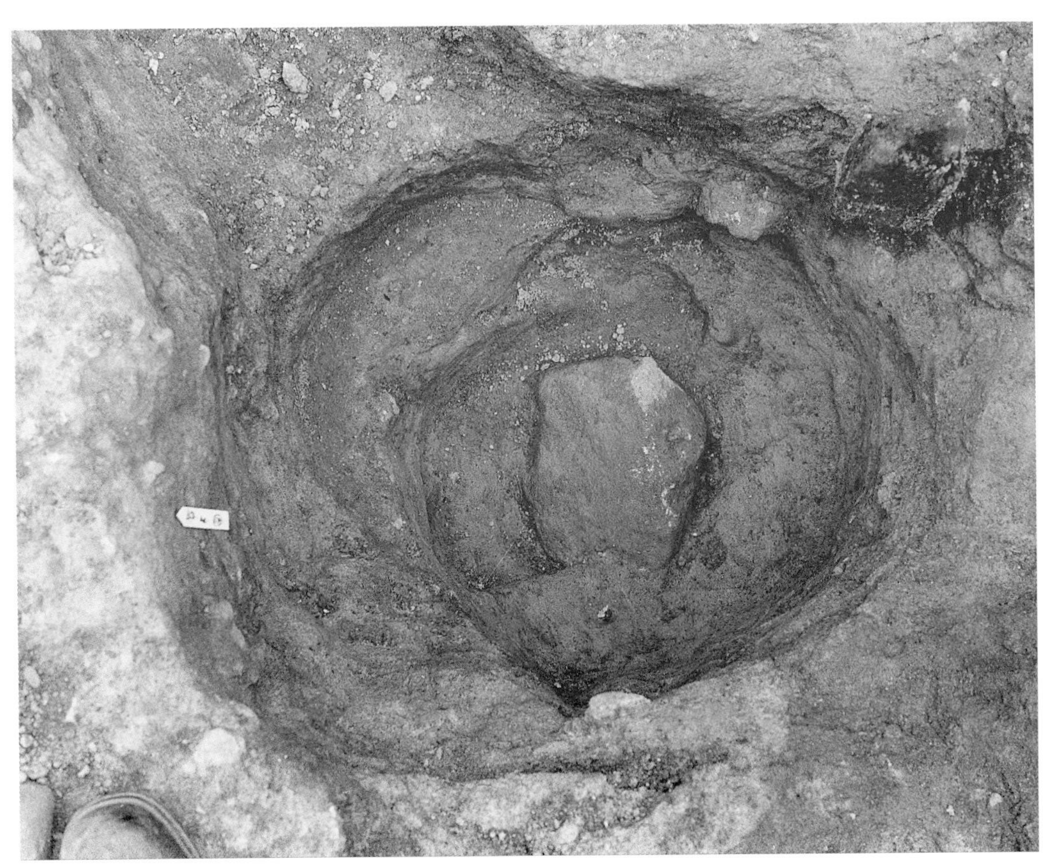

Plate 29   Southern Building: post-pit 1138 excavated

Plate 30   Southern Building: semi-circular socket as exposed, with altered top-step of ashpit visible at lower right (2 metre scale)

Plate 31   Southern Building: iron plates in final floor, from the north-east

Plate 32   Eastern Building: floor from above

Plate 33   Southern exterior: central section, from the east

Plate 34   South-east Building as excavated, from the north

Plate 35   North-east exterior: test trench to natural as completed, from the south

*Plate 36   North-east exterior: test trench to natural as completed; deposits abutting
buttress of furnace, from the north*

Plate 37   North-east exterior: test trench to natural; surface of redeposited natural, from the south

Plate 38   North-east exterior: Building 945 from the north

Plate 39   North-east exterior: Building 945, east side from the south

Plate 40   North-east exterior: Building 945, post impression at north-east corner

Plate 41   North-east exterior: Building 945, post-pit, brick wall infill, and post impression

Plate 42   North-east exterior: Building 1020, from the west

Plate 43   North-east exterior: Building 1020, north half from the south, with culvert 1030 at top right corner

Plate 44   West centre: timber slots in mortar surface. The right hand slot measures 0.96 x 0.22 m

Plate 45   West centre during excavation, showing bank of stratigraphy in northern bay (from north-west)

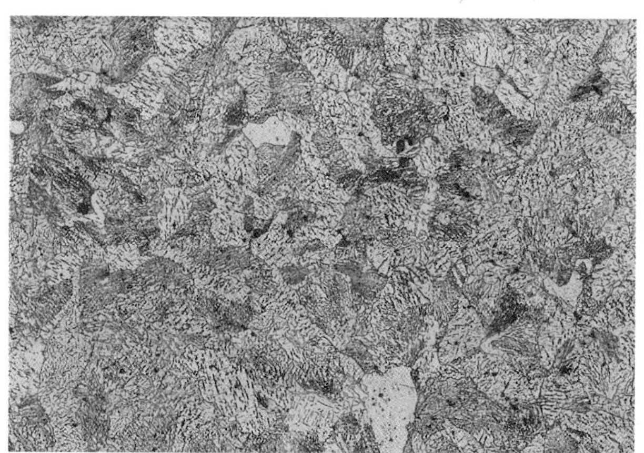

Plate 46   Sample 87/185 (Magnification x200). Coarse pearlite (dark etching) and proeutectoid ferrite (white). Note distribution of small inclusions and contrasts with 'typical' wrought iron slag inclusion size and distribution in Plate 47

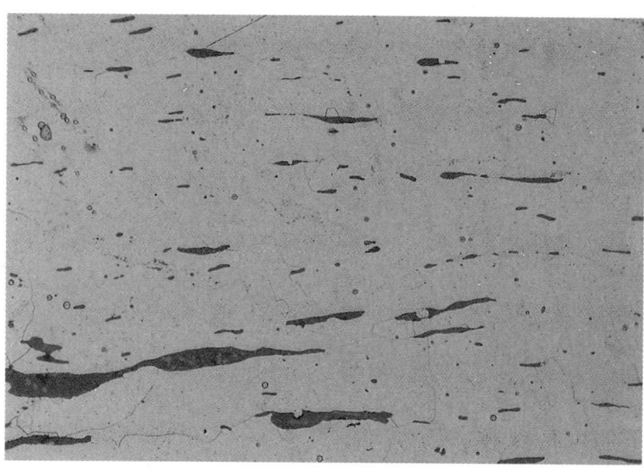

Plate 47   Sample 88/094 (Magnification x200). Coarse pearlite (dark etching) and proeutectoid ferrite (white). Austenite grain size can be observed

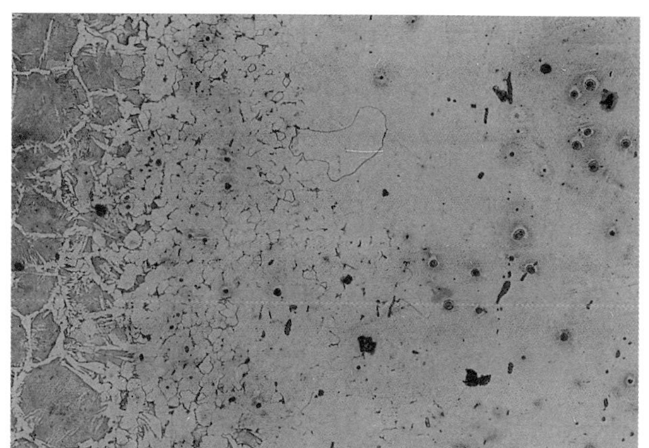

Plate 48   Sample 91/590 (Magnification x400). Typical wrought iron mircostructure ferrite grains (white) and orientated slag inclusions

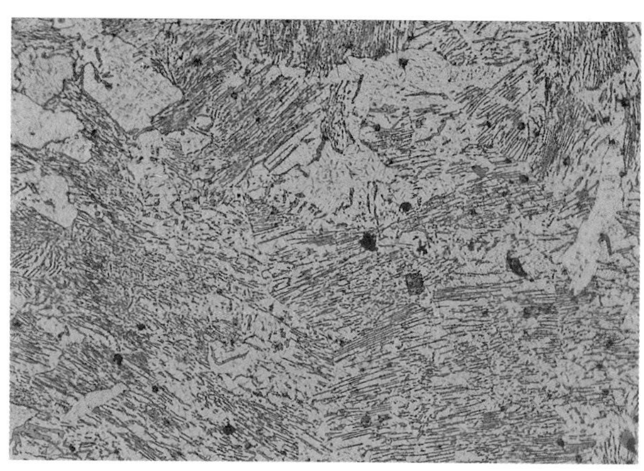

Plate 49   Sample 91/787 (Magnification x200). This sample shows a hetergeneous microstructure from ferrite grains (white) to pearlite (dark etching) plus proeutectoid grain boundary ferrite. Note slag inclusion pattern is similar to blister steel bars (Plates 45, 46) rather than the wrought iron sample (Plate 47)

Index